γ Gamma

Single and Multiple-Digit Multiplication

Student Workbook

1-888-854-MATH (6284) - mathusee.com
sales@mathusee.com

Gamma Student Workbook: Single and Multiple-Digit Multiplication
©2012 Math-U-See, Inc.
Published and distributed by Demme Learning

All rights reserved. No part of this book may be reproduced, stored in a retrieval system, or transmitted in any form by any means—electronic, mechanical, photocopying, recording, or otherwise—without prior written permission from Demme Learning.

mathusee.com

1-888-854-6284 or +1 717-283-1448 | demmelearning.com
Lancaster, Pennsylvania USA

ISBN 978-1-60826-068-3
Revision Code 1018-F

Printed in the United States of America by The P.A. Hutchison Company
7 8 9 10

For information regarding CPSIA on this printed material call: 1-888-854-6284 and provide reference #1018-08022022

Gamma

	LESSON PRACTICE			SYSTEMATIC REVIEW			
	A	B	C	D	E	F	TEST
1 Rectangle	☐	☐	☐	☐	☐	☐	☐
2 Multiply ×1, ×0	☐	☐	☐	☐	☐	☐	☐
3 Skip Count 2, 5, 10	☐	☐	☐	☐	☐	☐	☐
4 Multiply ×2	☐	☐	☐	☐	☐	☐	☐
5 Multiply ×10	☐	☐	☐	☐	☐	☐	☐
6 Multiply ×5	☐	☐	☐	☐	☐	☐	☐
Unit Test 1							☐
7 Area of Rectangle	☐	☐	☐	☐	☐	☐	☐
8 Solve Unknown	☐	☐	☐	☐	☐	☐	☐
9 Skip Count 9	☐	☐	☐	☐	☐	☐	☐
10 Multiply ×9	☐	☐	☐	☐	☐	☐	☐
11 Skip Count 3	☐	☐	☐	☐	☐	☐	☐
12 Multiply ×3	☐	☐	☐	☐	☐	☐	☐
13 Skip Count 6	☐	☐	☐	☐	☐	☐	☐
14 Multiply ×6	☐	☐	☐	☐	☐	☐	☐
Unit Test 2							☐
15 Skip Count 4	☐	☐	☐	☐	☐	☐	☐
16 Multiply ×4	☐	☐	☐	☐	☐	☐	☐
17 Skip Count 7	☐	☐	☐	☐	☐	☐	☐
18 Multiply ×7	☐	☐	☐	☐	☐	☐	☐
19 Skip Count 8	☐	☐	☐	☐	☐	☐	☐
20 Multiply ×8	☐	☐	☐	☐	☐	☐	☐
Unit Test 3							☐
21 Multiply Mult. Digits	☐	☐	☐	☐	☐	☐	☐
22 Round and Estimate	☐	☐	☐	☐	☐	☐	☐
23 2 Digit × 2 Digit	☐	☐	☐	☐	☐	☐	☐
24 2 Digit Regroup	☐	☐	☐	☐	☐	☐	☐
25 Mult. Digit Regroup	☐	☐	☐	☐	☐	☐	☐
26 Factors	☐	☐	☐	☐	☐	☐	☐
27 Millions	☐	☐	☐	☐	☐	☐	☐
28 Multiply Mult. Digits	☐	☐	☐	☐	☐	☐	☐
29 Prime Numbers	☐	☐	☐	☐	☐	☐	☐
30 Miles and Tons	☐	☐	☐	☐	☐	☐	☐
Unit Test 4							☐
Final Test							☐

APPLICATION AND ENRICHMENT PAGES

This edition of the *Gamma Student Workbook* includes extra activity pages titled "Application and Enrichment." You will find one enrichment page after the last systematic review page for each lesson. These activities are intended to do the following:

- Provide enjoyable practice of lesson concepts.
- Stimulate thinking by presenting concepts in different formats.
- Include activities suitable for a wide range of learning styles.
- Enrich learning with additional age-appropriate activities.
- Introduce new concepts that may be useful to students at this level.

The Application and Enrichment pages may be scheduled any time after the student has completed the corresponding lesson. Some activities may be challenging or require a new way of looking at a concept. Encourage students to think carefully for themselves, but do not hesitate to give them as much help as they need.

You can find helpful teaching tips and the solutions for the Application and Enrichment pages in the 2012 edition of the *Gamma Instruction Manual*.

LESSON PRACTICE 1A

Write the dimensions in the parentheses and the area in the oval. Then write the problem two ways beside the rectangle. The first one has been done for you.

1.

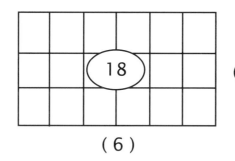

$6 \times 3 = 18$

$3 \times 6 = 18$

2.

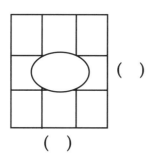

___ × ___ = ___

Write the problem only one way for a square. A square is a special kind of rectangle.

3.

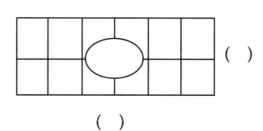

___ × ___ = ___

___ × ___ = ___

4.

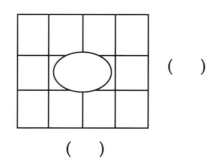

___ × ___ = ___

___ × ___ = ___

5.

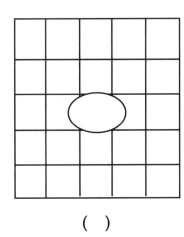
( )
( )

___ × ___ = ___

6.

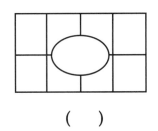
( )
( )

___ × ___ = ___

___ × ___ = ___

7. Build and draw the rectangle that is the same as:

8. Build and draw the rectangle that is the same as:

LESSON PRACTICE 1B

Write the dimensions in the parentheses and the area in the oval. Then write the problem beside the rectangle.

1.

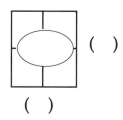 ( )

( )

____ × ____ = ____

2.

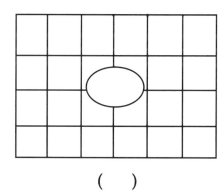 ( )

( )

____ × ____ = ____

____ × ____ = ____

3.

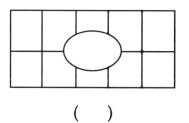 ( )

( )

____ × ____ = ____

____ × ____ = ____

4.

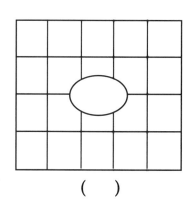 ( )

( )

____ × ____ = ____

____ × ____ = ____

GAMMA LESSON PRACTICE 1B

5.

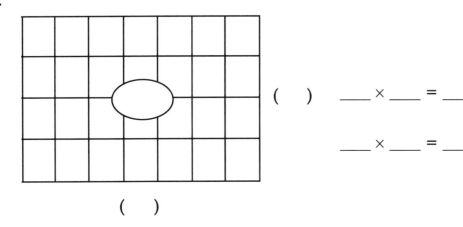

( )   ___ × ___ = ___

___ × ___ = ___

( )

6.

( )   ___ × ___ = ___

( )   ___ × ___ = ___

7. Build and draw the rectangle that is the same as:

8. Build and draw the rectangle that is the same as:

LESSON PRACTICE 1C

Write the dimensions in the parentheses and the area in the oval. Then write the problem beside the rectangle.

1.
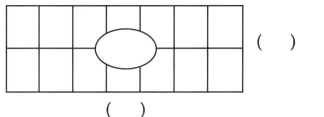
( ) ___ × ___ = ___

___ × ___ = ___

2.
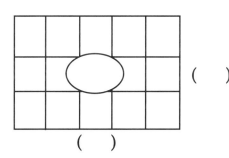
( ) ___ × ___ = ___

___ × ___ = ___

3.
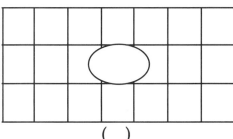
( ) ___ × ___ = ___

___ × ___ = ___

4.
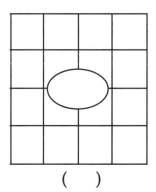
( ) ___ × ___ = ___

5.

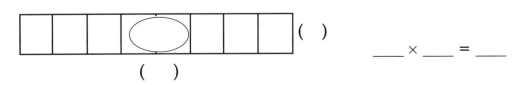

( )
( )

___ × ___ = ___

6.

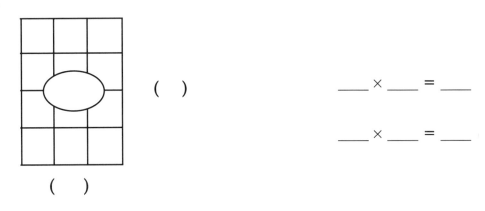

( )
( )

___ × ___ = ___

___ × ___ = ___

7. Build and draw the rectangle that is the same as:

☐☐ + ☐☐ + ☐☐ + ☐☐ + ☐☐

8. Build and draw the rectangle that is the same as:

☐☐☐☐ + ☐☐☐☐

SYSTEMATIC REVIEW

1D

Write the dimensions in the parentheses and the area in the oval. Then write the problem two ways beside the rectangle.

1. 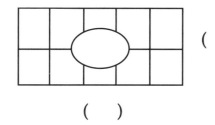 ( )

   ( )

   ___ × ___ = ___

   ___ × ___ = ___

2. 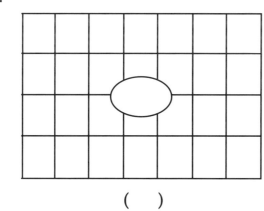 ( )

   ( )

   ___ × ___ = ___

   ___ × ___ = ___

Add.

3.   3
   + 1

4.   4
   + 2

5.   2
   + 8

6.   9
   + 5

SYSTEMATIC REVIEW 1D

7. 8
   +6

8. 5
   +5

9. 7
   +6

10. 4
   +5

Subtract.

11. 10
    − 1

12. 8
    − 2

13. 15
    − 9

14. 16
    − 8

15. 8
    − 4

16. 7
    − 3

17. 9
    − 6

18. 11
    − 7

SYSTEMATIC REVIEW

Write the dimensions in the parentheses and the area in the oval. Then write the problem two ways beside the rectangle.

1. 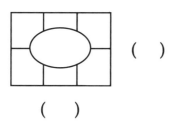   ( )

   ( )

   ___ × ___ = ___

   ___ × ___ = ___

2. 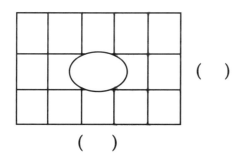   ( )

   ( )

   ___ × ___ = ___

   ___ × ___ = ___

Add.

3. 6
   + 1

4. 5
   + 2

5. 2
   + 3

6. 9
   + 7

7. 8
   + 4

8. 6
   + 6

SYSTEMATIC REVIEW 1E

9. 5
   + 6

10. 4
    + 7

Subtract.

11. 8
    − 1

12. 5
    − 2

13. 1 8
    − 9

14. 1 4
    − 7

15. 1 5
    − 8

16. 8
    − 5

17. 9
    − 8

18. 1 2
    − 4

SYSTEMATIC REVIEW

Write the dimensions in the parentheses and the area in the oval. Then write the problem beside the rectangle.

1. 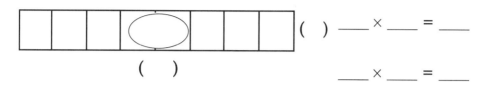 ( ) ___ × ___ = ___

   ___ × ___ = ___

2.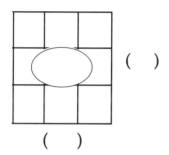

   ___ × ___ = ___

Add.

3. 7
   + 1

4. 6
   + 2

5. 2
   + 9

6. 9
   + 4

7. 8
   + 3

8. 4
   + 4

9. 7
  + 8

10. 5
  + 7

Subtract.

11. 3
  − 1

12. 9
  − 2

13. 1 6
  − 9

14. 1 3
  − 8

15. 1 2
  − 6

16. 7
  − 4

17. 1 3
  − 7

18. 1 2
  − 3

APPLICATION AND ENRICHMENT

Here is a chance to build some rectangles of your own.

1. Choose 4 three bars and build a rectangle.

   What are the dimensions of your rectangle?

   ____ × ____ and ____ × ____

   What is the area of your rectangle?

2. Choose 3 seven bars and build a rectangle.

   What are the dimensions of your rectangle?

   ____ × ____ and ____ × ____

   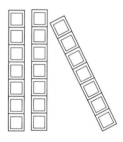

   What is the area of your rectangle?

3. Choose 6 ten bars and build a rectangle.

   What are the dimensions of your rectangle?

   ____ × ____ and ____ × ____

   What is the area of your rectangle?

APPLICATION AND ENRICHMENT 1G

Here is a chance to build some rectangles of your own.

4. Use four bars to build a rectangle that has an area of 12.

   How many four bars did you use?

   What are the dimensions of your rectangle?

   ____ × ____ and ____ × ____

5. Now use three bars to build a rectangle with an area of 12.

   How many three bars did you use?

   What are the dimensions of your rectangle?

   ____ × ____ and ____ × ____

6. Next use six bars to build a rectangle with an area of 12.

   How many six bars did you use?

   What are the dimensions of your rectangle?

   ____ × ____ and ____ × ____

7. Can rectangles with different dimensions have the same area?

LESSON PRACTICE 2A

Find the answer by multiplying.

1. $1 \times 0 =$ ____

2. $0 \times 3 =$ ____

3. $4 \times 0 =$ ____

4. $0 \times 6 =$ ____

5. $(9)(0) =$ ____

6. $(0)(2) =$ ____

7. $(0)(5) =$ ____

8. $(8)(0) =$ ____

9. $7 \cdot 0 =$ ____

10. $1 \cdot 1 =$ ____

11. $8 \cdot 1 =$ ____

12. $1 \cdot 2 =$ ____

13. $3 \times 1 =$ ____

14. $(1)(5) =$ ____

15. $(7)(1) =$ ____

16. $1 \cdot 4 =$ ____

17. 9
    $\times 1$

18. 1
    $\times 6$

19. 1
    $\times 0$

20. 0
    $\times 5$

GAMMA LESSON PRACTICE 2A

LESSON PRACTICE 2A

Write the dimensions in the parentheses and the area in the oval. Then write the problem two ways.

21.

___ × ___ = ___

___ × ___ = ___

22. How much is zero counted six times?

23. How much is eight counted one time?

24. Sally has one flower pot on her table. In the pot are three daffodils. How many flowers are on the table?

25. There are nine books on the table. Tom put one bookmark in each book. How many bookmarks did he need?

Note: See lesson 2 in the instruction manual for tips on teaching word problems.

LESSON PRACTICE 2B

Find the answer by multiplying.

1. $5 \times 0 =$ _____

2. $0 \times 1 =$ _____

3. $6 \times 1 =$ _____

4. $1 \times 9 =$ _____

5. $(1)(4) =$ _____

6. $(7)(1) =$ _____

7. $(1)(5) =$ _____

8. $(3)(1) =$ _____

9. $1 \cdot 2 =$ _____

10. $8 \cdot 1 =$ _____

11. $1 \cdot 1 =$ _____

12. $7 \cdot 0 =$ _____

13. $8 \times 0 =$ _____

14. $(0)(5) =$ _____

15. $(0)(2) =$ _____

16. $0 \cdot 9 =$ _____

17. 0
    $\times\ 1$
    ─────

18. 3
    $\times\ 0$
    ─────

19. 0
    $\times\ 4$
    ─────

20. 6
    $\times\ 0$
    ─────

GAMMA LESSON PRACTICE 2B

LESSON PRACTICE 2B

Write the dimensions in the parentheses and the area in the oval. Then write the problem two ways.

21.

    ( )   ___ × ___ = ___

    ( )   ___ × ___ = ___

22. How much is zero counted one time?

23. How much is 10 counted one time?

24. Mother said that Todd had to do zero chores each day during vacation. Vacation lasted for five days. How many chores did Todd have to do?

25. Every person needs one plate. There are six people in the family. How many plates are needed?

LESSON PRACTICE   2C

Find the answer by multiplying.

1. $6 \times 0 =$ _____

2. $0 \times 2 =$ _____

3. $5 \times 1 =$ _____

4. $1 \times 8 =$ _____

5. $(1)(7) =$ _____

6. $(6)(1) =$ _____

7. $(1)(2) =$ _____

8. $(9)(1) =$ _____

9. $1 \cdot 5 =$ _____

10. $4 \cdot 1 =$ _____

11. $1 \cdot 0 =$ _____

12. $6 \cdot 0 =$ _____

13. $4 \times 0 =$ _____

14. $(0)(3) =$ _____

15. $(0)(8) =$ _____

16. $0 \cdot 2 =$ _____

17. 3
    $\times\ 1$

18. 7
    $\times\ 0$

19. 1
    $\times\ 4$

20. 1
    $\times\ 8$

GAMMA   LESSON PRACTICE 2C

LESSON PRACTICE 2C

Write the dimensions in the parentheses and the area in the oval. Then write the problem two ways.

21.
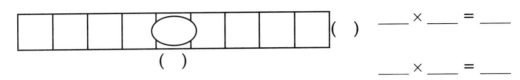

____ × ____ = ____

____ × ____ = ____

22. How much is zero counted three times?

23. How much is nine counted one time?

24. Everyone in the family has one nose. There are eight people in the family. How many noses are there?

25. Paul has seven apple trees in his yard. If every tree grows zero apples, how many apples will grow?

SYSTEMATIC REVIEW

2D

Find the answer by multiplying.

1. $1 \times 9 =$ _____

2. $3 \times 1 =$ _____

3. $7 \times 0 =$ _____

4. $9 \times 0 =$ _____

5. $(1)(0) =$ _____

6. $(1)(6) =$ _____

7. $(1)(5) =$ _____

8. $(1)(1) =$ _____

9. $0 \cdot 2 =$ _____

10. $0 \cdot 3 =$ _____

11. $0 \cdot 1 =$ _____

12. $7 \cdot 1 =$ _____

13. $\begin{array}{r} 4 \\ \times\, 1 \\ \hline \end{array}$

14. $\begin{array}{r} 2 \\ \times\, 1 \\ \hline \end{array}$

15. $\begin{array}{r} 0 \\ \times\, 8 \\ \hline \end{array}$

16. $\begin{array}{r} 6 \\ \times\, 0 \\ \hline \end{array}$

SYSTEMATIC REVIEW 2D

Add or subtract.

17. 5
    + 7

18. 3
    + 4

19. 7
    − 5

20. 5
    + 4

21. 4
    − 0

22. 6
    + 6

23. 6
    + 3

24. 1 5
    − 9

25. How many is zero counted seven times?

26. Each cage holds one bunny. How many bunnies are in eight cages?

SYSTEMATIC REVIEW

Find the answer by multiplying.

1. $0 \times 6 =$ _____

2. $8 \times 0 =$ _____

3. $2 \times 1 =$ _____

4. $1 \times 4 =$ _____

5. $(0)(2) =$ _____

6. $(0)(3) =$ _____

7. $(1)(0) =$ _____

8. $(7)(1) =$ _____

9. $8 \times 1 =$ _____

10. $(0)(5) =$ _____

11. $4 \cdot 1 =$ _____

12. $5 \cdot 0 =$ _____

13. 0
    $\underline{\times\,1}$

14. 6
    $\underline{\times\,1}$

15. 5
    $\underline{\times\,1}$

16. 1
    $\underline{\times\,1}$

SYSTEMATIC REVIEW 2E

Add or subtract.

17.   5
    + 3

18.   8
    + 6

19. 1 5
    − 7

20.   7
    + 3

21. 1 2
    − 7

22.   7
    + 6

23.   6
    + 5

24. 1 1
    − 3

25. How many is three counted one time?

26. Mom wanted each of her children to have a new coat. She has eight children. How many coats must she buy?

SYSTEMATIC REVIEW

Find the answer by multiplying.

1. $1 \times 10 =$ _____

2. $4 \times 1 =$ _____

3. $6 \times 0 =$ _____

4. $1 \times 7 =$ _____

5. $0 \cdot 8 =$ _____

6. $(9)(1) =$ _____

7. $5 \cdot 1 =$ _____

8. $(7)(0) =$ _____

9.   6
  × 1

10.   1
  × 1

11.   0
  × 3

12.   10
  × 0

Add or subtract.

13.   16
  − 8

14.   4
  + 5

SYSTEMATIC REVIEW 2F

Add or subtract.

15. 7
   − 4

16. 8
   + 2

17. 6
   + 4

18. 1 2
   − 5

19. 5
   + 6

20. 3
   + 6

21. 1 3
   − 5

22. 8
   − 0

23. 3
   + 9

24. 1 7
   − 8

25. How many is zero counted two times?

26. The apple pie recipe calls for zero cans of tomatoes. How many cans of tomatoes are needed for four apple pies?

Multiplication Facts Sheet

0×0	0×1	0×2	0×3	0×4	0×5	0×6	0×7	0×8	0×9	0×10
1×0	1×1	1×2	1×3	1×4	1×5	1×6	1×7	1×8	1×9	1×10
2×0	2×1	2×2	2×3	2×4	2×5	2×6	2×7	2×8	2×9	2×10
3×0	3×1	3×2	3×3	3×4	3×5	3×6	3×7	3×8	3×9	3×10
4×0	4×1	4×2	4×3	4×4	4×5	4×6	4×7	4×8	4×9	4×10
5×0	5×1	5×2	5×3	5×4	5×5	5×6	5×7	5×8	5×9	5×10
6×0	6×1	6×2	6×3	6×4	6×5	6×6	6×7	6×8	6×9	6×10
7×0	7×1	7×2	7×3	7×4	7×5	7×6	7×7	7×8	7×9	7×10
8×0	8×1	8×2	8×3	8×4	8×5	8×6	8×7	8×8	8×9	8×10
9×0	9×1	9×2	9×3	9×4	9×5	9×6	9×7	9×8	9×9	9×10
10×0	10×1	10×2	10×3	10×4	10×5	10×6	10×7	10×8	10×9	10×10

MULTIPLICATION FACTS SHEET   GAMMA INSTRUCTION MANUAL, PUBLISHED BY MATH-U-SEE, 2012

APPLICATION AND ENRICHMENT

2G

Tom's dad let him drive the car zero times on Sunday.
Write 0 in the square for Sunday.

Sun	Mon	Tue	Wed	Thu	Fri	Sat

For seven days, Tom drove the car zero times every day.
Write a zero in each square on the calendar.

1. How many times did Tom drive the car that week?

    $0 \times 7 = $ _____ times

Sam has several pets. He has five little homes for his pets, and every home has one pet. Draw one pet in each home.

2. How many pets does Sam have altogether?

    $1 \times 5 = $ _____ pets

APPLICATION AND ENRICHMENT 2G

Write the answers to the problems in each box. Are the answers the same? Circle yes or no.

3.
$1 \times 4 = \underline{}$

$(4)(1) = \underline{}$

yes

no

4.
$4 \times 1 = \underline{}$

$4 - 1 = \underline{}$

yes

no

5.
$0 \cdot 3 = \underline{}$

$0 + 0 + 0 = \underline{}$

yes

no

6.
$1 \times 3 = \underline{}$

$1 + 3 = \underline{}$

yes

no

7. Write one times four several different ways. Remember that multiplication is also fast adding.

_____

LESSON PRACTICE 3A

Skip count by 2, by 10, and by 5. Write the correct numbers in the squares with the lines.

1.

2.

GAMMA LESSON PRACTICE 3A

LESSON PRACTICE 3A

3.

				10
				15
				50

4.

				5

Skip count and write the numbers.

5. 2, _____, _____, _____, _____, 12, _____, _____, _____, _____

6. 5, _____, 15, _____, _____, _____, _____, _____, _____, _____

7. 10, _____, _____, _____, _____, _____, _____, _____, _____, 100

LESSON PRACTICE 3B

Skip count by 2, by 10, and by 5. Write the correct numbers in the squares with the lines.

1.

2.

GAMMA LESSON PRACTICE 3B

LESSON PRACTICE 3B

3.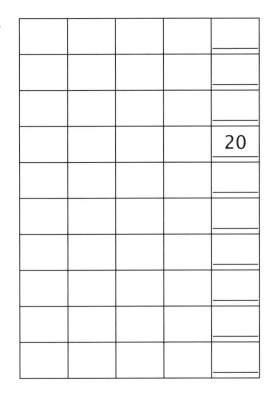

4.

Skip count and write the numbers.

5. 2, ____, ____, ____, ____, ____, ____, ____, ____, 20

6. 5, ____, ____, ____, ____, 30, ____, ____, ____, ____

7. 10, ____, ____, ____, ____, 60, ____, ____, ____, ____

LESSON PRACTICE 3C

Skip count by 2, by 10, and by 5. Write the correct numbers in the squares with the lines.

1.

	___
	4
	___
	___
	___
	___
	14
	___
	___
	___

2.

									___
									___
									___
									40
									___
									___
									___
									___
									___
									___

GAMMA LESSON PRACTICE 3C

LESSON PRACTICE 3C

3.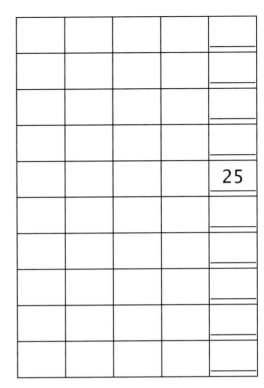

4.

Skip count and write the numbers.

5. 2, ____, ____, ____, ____, ____, ____, ____, 18, ____

6. 5, ____, ____, ____, ____, ____, ____, 40, ____, ____

7. 10, ____, 30, ____, ____, ____, ____, ____, ____, ____

SYSTEMATIC REVIEW

Skip count and write the numbers.

1. 5, ___, ___, ___, ___, ___, ___, ___, ___, ___

2. 10, ___, ___, ___, ___, ___, ___, ___, ___, ___

3. 2, ___, ___, ___, ___, ___, ___, ___, ___, ___

Find the answer by multiplying.

4. 1 • 7 = ___

5. 6 × 1 = ___

6. 10 • 0 = ___

7. (0)(2) = ___

8. 3
   × 1
   ___

9. 8
   × 1
   ___

10. 0
    × 4
    ___

11. 0
    × 0
    ___

Add or subtract.

12. 7
    + 8
    ___

13. 1 4
    − 5
    ___

14. 12
     − 4

15.   8
     + 9

16. How many is zero counted nine times?

17. Each flower has five petals. Jen needs 35 dried petals for her art project. How many flowers must she pick? (Use your knowledge of skip counting.)

18. Eunice gave one cookie to each of her seven children. How many cookies did she give them altogether?

19. Hannah read five chapters of her book yesterday and six chapters today. How many chapters has she read?

20. Each of the children in the class owned zero rocket ships. There are seven boys and nine girls. How many rocket ships did they own altogether? (This is a simple two-step problem. Watch for more two-step problems in this book.)

SYSTEMATIC REVIEW

3E

Skip count and write the numbers.

1. ___, ___, 15, ___, ___, ___, ___, ___, ___, ___

2. ___, 4, ___, ___, ___, ___, ___, ___, ___, ___

3. ___, 20, ___, ___, ___, ___, ___, ___, ___, ___

Find the answer by multiplying.

4. $5 \cdot 1 =$ ___

5. $6 \times 0 =$ ___

6. $1 \cdot 10 =$ ___

7. $(9)(1) =$ ___

8. 0
   $\times\,3$

9. 1
   $\times\,1$

10. 0
    $\times\,8$

11. 4
    $\times\,1$

Add or subtract.

12. 8
    $-\,4$

13. 1 0
    $+\,3$

14. 13
    − 6

15.   5
    + 7

16. How many is zero counted zero times?

17. Each bag contains 10 oranges. I need 50 oranges. How many bags must I buy? (skip counting)

18. Julia climbed the mountain one time each day for a week. There are seven days in a week. How many times did Julia climb the mountain?

19. David had 11 snacks in his backpack. He ate six snacks. How many snacks are left?

20. My mother gave me $7, and my father gave me $3. Then I subtracted $1 for saving. How much money do I have left?

SYSTEMATIC REVIEW

Skip count and write the numbers.

1. ____, ____, ____, ____, ____, ____, ____, ____, ____, 100

2. ____, ____, ____, ____, ____, ____, ____, ____, ____, 50

3. ____, ____, ____, ____, ____, ____, ____, ____, ____, 20

Find the answer by multiplying.

4. $1 \cdot 0 =$ ____

5. $2 \times 1 =$ ____

6. $5 \cdot 0 =$ ____

7. $(8)(1) =$ ____

8. 6
   $\underline{\times\,1}$

9. 10
   $\underline{\times\,0}$

10. 7
    $\underline{\times\,1}$

11. 9
    $\underline{\times\,0}$

Add or subtract.

12. 6
    $\underline{+\,7}$

13. 18
    $\underline{-\,9}$

Add or subtract.

14. 13
    − 4

15.   8
    + 5

16. What number equals five counted one time?

17. Kim must knit two mittens for each child. How many mittens must she knit for nine children? (skip counting)

18. Ten children are in the park. Each child is flying one kite. How many kites are being flown?

19. Peter picked nine apples. He found that two of the apples were rotten and that two were wormy. How many good apples did Peter have left?

20. Timothy read zero stories a day for five days. Then he read one story a day for six days. How many stories has he read so far?

APPLICATION AND ENRICHMENT

Skip count by five. Start at the star and connect the dots all the way to 100. Use the picture to practice skip counting by five.

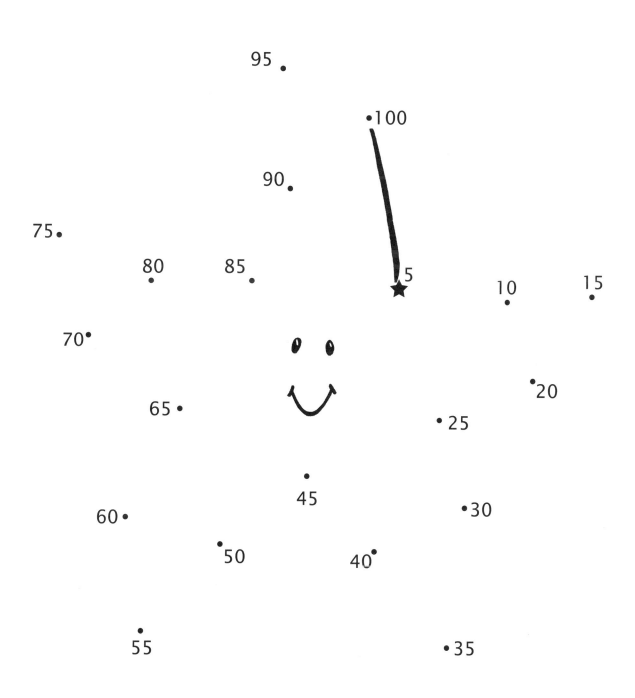

Skip count by two. Start at the star and connect the dots all the way to 100. Use the picture to practice skip counting by two.

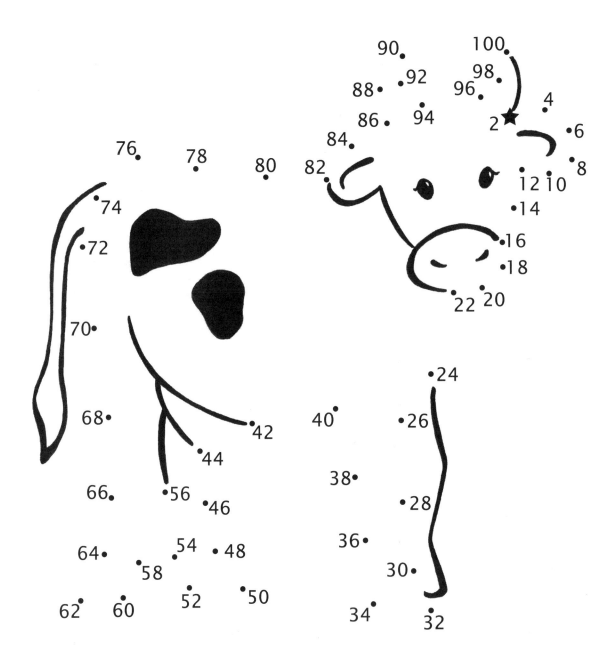

LESSON PRACTICE 4A

Find the answer by multiplying.

1. $1 \times 2 = $ _____

2. $2 \times 3 = $ _____

3. $(2)(5) = $ _____

4. $(2)(7) = $ _____

5. $(9)(2) = $ _____

6. $(2)(2) = $ _____

7. $2 \cdot 10 = $ _____

8. $8 \cdot 2 = $ _____

9.   4
   $\times\ 2$

10.   7
    $\times\ 2$

11.   2
    $\times\ 6$

12.   2
    $\times\ 0$

Write the factors in the parentheses and the product in the oval. Write the problem two ways.

13.

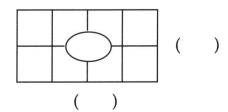

( )

( )

_____ × _____ = _____

_____ × _____ = _____

LESSON PRACTICE 4A

Skip count and write the missing numbers. Then fill in the missing factors under the lines.

14–15.

$$\frac{0}{2\times 0} \quad \frac{2}{2\times 1} \quad \frac{}{2\times 2} \quad \frac{6}{2\times \underline{}} \quad \frac{}{2\times \underline{}} \quad \frac{}{2\times 5}$$

$$\frac{12}{2\times \underline{}} \quad \frac{}{2\times 7} \quad \frac{}{2\times 8} \quad \frac{}{2\times 9} \quad \frac{20}{2\times 10}$$

Multiply the quarts by 2 to find the number of pints.

16.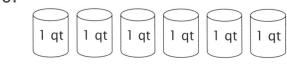

    $6 \times 2 = $ _____

17. What is another way to write 2×8? _____

18. Two counted 10 times equals _____ .

19. Each picture book had 10 pages. Hope read two of the books to her little brother. How many pages did Hope read?

20. Jean bought seven quarts of soda. How many pints of soda did she buy?

LESSON PRACTICE 4B

Find the answer by multiplying.

1. 2 × 2 = _____

2. 2 × 6 = _____

3. (2)(8) = _____

4. (9)(2) = _____

5. (2)(7) = _____

6. (4)(2) = _____

7. 1 • 2 = _____

8. 3 • 2 = _____

9. 0
   × 2
   ___

10. 5
    × 2
    ___

11. 2
    × 8
    ___

12. 1 0
    × 2
    ___

Write the factors in the parentheses and the product in the oval. Write the problem two ways.

13.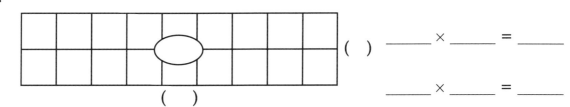

LESSON PRACTICE 4B

Skip count and write the missing numbers. Then fill in the missing factors under the lines.

14–15.

$$\frac{0}{2\cdot 0} \quad \frac{2}{2\cdot \_\_} \quad \frac{\_\_}{2\cdot 2} \quad \frac{\_\_}{2\cdot 3} \quad \frac{8}{2\cdot \_\_} \quad \frac{\_\_}{2\cdot 5}$$

$$\frac{\_\_}{2\cdot 6} \quad \frac{14}{2\cdot \_\_} \quad \frac{\_\_}{2\cdot 8} \quad \frac{\_\_}{2\cdot 9} \quad \frac{\_\_}{2\cdot 10}$$

Multiply the quarts by 2 to find the number of pints.

16.

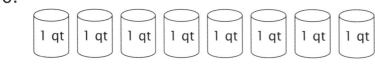

$2 \times 8 = $ _____

17. What is another way to write 2×6?

18. Two counted seven times equals _____ .

19. Amanda has three sets of twins. How many children is that?

20. George made eight quarts of punch for the party. How many pints of punch did he make?

LESSON PRACTICE 4C

Find the answer by multiplying.

1. 9 × 2 = _____
2. 2 × 4 = _____

3. (2)(7) = _____
4. (8)(2) = _____

5. (5)(2) = _____
6. (1)(2) = _____

7. 2 • 6 = _____
8. 10 • 2 = _____

9.   2
   × 2

10.   3
    × 2

11.   2
    × 9

12.   0
    × 2

Write the factors in the parentheses and the product in the oval. Write the problem in the blanks.

13.

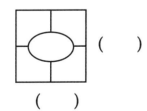

( )

( )

_____ × _____ = _____

LESSON PRACTICE 4C

Skip count and write the missing numbers. Then fill in the missing factors under the lines.

14-15.

$$\frac{0}{(2)(0)} \quad \frac{2}{2(\_\_)} \quad \frac{\_\_}{(2)(2)} \quad \frac{6}{2(\_\_)} \quad \frac{\_\_}{(2)(4)} \quad \frac{10}{(2)(\_\_)}$$

$$\frac{\_\_}{(2)(6)} \quad \frac{\_\_}{2(\_\_)} \quad \frac{\_\_}{(2)(8)} \quad \frac{18}{2(\_\_)} \quad \frac{\_\_}{(2)(10)}$$

Multiply the quarts by 2 to find the number of pints.

16.

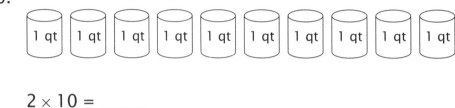

2 × 10 = _____

17. What is another way to write 2 × 7? _____

18. Two counted two times equals _____ .

19. Susan made nine quarts of formula. How many pints of formula did she make?

20. Mary keeps two rabbits in each cage. She has six cages. How many rabbits does she have?

SYSTEMATIC REVIEW

Find the answer by multiplying.

1. $5 \cdot 2 =$ _____
2. $2 \times 6 =$ _____

3. $9 \cdot 2 =$ _____
4. $(2)(8) =$ _____

5. $\begin{array}{r} 1 \\ \times\,3 \\ \hline \end{array}$
6. $\begin{array}{r} 2 \\ \times\,7 \\ \hline \end{array}$

7. $\begin{array}{r} 0 \\ \times\,6 \\ \hline \end{array}$
8. $\begin{array}{r} 4 \\ \times\,2 \\ \hline \end{array}$

9. $3 \times 2 =$ _____

   $2 \times 3 =$ _____

10. $10 \times 2 =$ _____

    $2 \times 10 =$ _____

11. $2 \times 1 =$ _____

    $1 \times 2 =$ _____

12. $0 \times 2 =$ _____

    $2 \times 0 =$ _____

Skip count and write the numbers.

13. 2, ____, ____, ____, ____, ____, ____, ____, ____, ____

SYSTEMATIC REVIEW 4D

Add or subtract.

14. 12
    − 4

15.   9
    + 8

16. 15
    − 7

17.   5
    + 4

QUICK REVIEW

Write numbers with place-value notation to show how many hundreds, tens, and units they represent.

Example 1
143 = 100 + 40 + 3

18. 542 = ____ + ____ + ____

19. 163 = ____ + ____ + ____

20. What is two counted seven times?

21. Andy has two pints of milk. Anne has two times as much milk. How much milk does Anne have?

22. Bob and Sue have five girls and three boys. How many children do they have?

They gave each of their children a bicycle for Christmas. How many wheels were on all the bicycles?

SYSTEMATIC REVIEW 4E

Find the answer by multiplying.

1. 0 • 2 = _____
2. 5 × 2 = _____

3. 2 • 2 = _____
4. (2)(4) = _____

5.   2
  × 3
  ‾‾‾

6.   9
  × 1
  ‾‾‾

7.   2
  × 6
  ‾‾‾

8.   1 0
  × 2
  ‾‾‾

9. 7 × 2 = _____

   2 × 7 = _____

10. 8 × 2 = _____

    2 × 8 = _____

11. 5 × 1 = _____

    1 × 5 = _____

12. 2 × 6 = _____

    6 × 2 = _____

Skip count and write the numbers.

13. 10, ____, ____, ____, ____, ____, ____, ____, ____, ____

Add or subtract.

14. 16
    − 8

15. 5
    + 3

16. 18
    − 9

17. 7
    + 5

Rewrite using place-value notation.

18. 351 = _____ + _____ + _____

19. 249 = _____ + _____ + _____

20. What is seven counted two times? (Remember that you can change the order of the question.)

21. An octopus has eight arms. Two octopuses bought mittens. If an octopus wears one mitten on each arm, how many mittens did they need to buy altogether?

22. Danny picked up five eggs in each hand and then dropped three eggs. How many whole eggs are left?

SYSTEMATIC REVIEW

Find the answer by multiplying.

1. 3 · 2 = _____
2. 2 × 10 = _____

3. 8 · 2 = _____
4. (1)(7) = _____

5. 2
   × 3
   ___

6. 2
   × 6
   ___

7. 4
   × 2
   ___

8. 0
   × 9
   ___

9. 5 × 2 = _____

   2 × 5 = _____

10. 7 × 2 = _____

    2 × 7 = _____

11. 9 × 2 = _____

    2 × 9 = _____

12. 4 × 1 = _____

    1 × 4 = _____

Skip count and write the numbers.

13. 5, ___, ___, ___, ___, ___, ___, ___, ___, ___

SYSTEMATIC REVIEW 4F

Add or subtract.

14. 16
    − 9

15.   7
    + 7

16.   9
    − 4

17.   5
    + 6

Rewrite using place-value notation.

18. 131 = ____ + ____ + ____

19. 475 = ____ + ____ + ____

20. Each of the two elephants ate 10 peanuts. How many peanuts did they eat altogether?

21. Lauren needs three quarts of ice cream, but it comes only in pint cartons. How many pints should she buy?

22. Edith knitted two hats for each of her two nieces. How many hats did Edith knit for them?

She knitted three hats apiece for her two nephews. How many did she knit for them?

How many hats did Edith knit in all?

APPLICATION AND ENRICHMENT

A graph that uses pictures to show information is called a *pictograph*.

Mr. Smith sold milk at his corner store. Here is how much he sold for the last five days: Monday-6 quarts, Tuesday-4 quarts, Wednesday-6 quarts, Thursday-8 quarts, Friday-10 quarts.

Color the right number of quart milk cartons to show how much milk Mr. Smith sold each day. Use the pictograph to answer the questions.

1. On what day did Mr. Smith sell the most quarts of milk?

2. On what day did he sell the fewest quarts of milk?

3. On which two days was the same amount of milk sold?

4. How many pints of milk were sold on Thursday? (multiply by 2)

Use the words and clues to fill in the crossword puzzle.

area
factors
pints
product
quart
rectangle
square

Across

1. When we multiply, the answer is called the _____ .

3. Two pints make one _____ .

4. A _____ has four sides and four square corners.

6. In $2 \times 6 = 12$, the 2 and the 6 are the _____ .

Down

1. Four quarts is the same as eight _____ .

2. If all four sides of a rectangle are the same length, it is also a _____ .

5. Multiply the dimensions of a rectangle to find the _____ .

LESSON PRACTICE 5A

Find the answer by multiplying.

1. $10 \times 0 =$ _____

2. $5 \times 10 =$ _____

3. $10 \times 2 =$ _____

4. $6 \times 10 =$ _____

5. $(10)(10) =$ _____

6. $(10)(3) =$ _____

7. $10 \cdot 9 =$ _____

8. $10 \cdot 7 =$ _____

9. $\quad 10$
   $\underline{\times\ 2}$

10. $\quad 10$
    $\underline{\times\ 5}$

11. $\quad 10$
    $\underline{\times\ 1}$

12. $\quad 10$
    $\underline{\times\ 3}$

13. $10 \times 7 =$ _____

    $7 \times 10 =$ _____

14. $4 \times 10 =$ _____

    $10 \times 4 =$ _____

15. $10 \times 6 =$ _____

    $6 \times 10 =$ _____

16. $10 \times 3 =$ _____

    $3 \times 10 =$ _____

LESSON PRACTICE 5A

Color all the boxes that have a number you would say when skip counting by 10. Notice the pattern.

17.

0	1	2	3	4	5	6	7	8	9
10	11	12	13	14	15	16	17	18	19
20	21	22	23	24	25	26	27	28	29
30	31	32	33	34	35	36	37	38	39
40	41	42	43	44	45	46	47	48	49
50	51	52	53	54	55	56	57	58	59
60	61	62	63	64	65	66	67	68	69
70	71	72	73	74	75	76	77	78	79
80	81	82	83	84	85	86	87	88	89
90	91	92	93	94	95	96	97	98	99

18. How many pennies or cents are the same as four dimes?

19. Ten counted nine times equals _____ .

20. Ten cars went by the house every hour. How many cars went by in six hours?

LESSON PRACTICE 5B

Find the answer by multiplying.

1. $10 \times 8 =$ _____
2. $1 \times 10 =$ _____

3. $10 \times 9 =$ _____
4. $0 \times 10 =$ _____

5. $(10)(5) =$ _____
6. $(10)(4) =$ _____

7. $10 \cdot 6 =$ _____
8. $10 \cdot 10 =$ _____

9. 10
   $\underline{\times\ 8}$

10. 10
    $\underline{\times\ 7}$

11. 10
    $\underline{\times\ 2}$

12. 10
    $\underline{\times\ 1}$

13. $10 \times 5 =$ _____
14. $8 \times 10 =$ _____

    $5 \times 10 =$ _____
    $10 \times 8 =$ _____

15. $10 \times 0 =$ _____
16. $10 \times 9 =$ _____

    $0 \times 10 =$ _____
    $9 \times 10 =$ _____

GAMMA LESSON PRACTICE 5B

LESSON PRACTICE 5B

Skip count and write the missing numbers. Then fill in the missing factors.

17.
$$\frac{0}{(10)(0)} \quad \frac{10}{(10)()} \quad \frac{}{(10)(2)} \quad \frac{30}{(10)()} \quad \frac{}{(10)(4)} \quad \frac{}{(10)()}$$

$$\frac{}{(10)(6)} \quad \frac{}{(10)()} \quad \frac{}{(10)(8)} \quad \frac{90}{(10)()} \quad \frac{}{(10)(10)}$$

18. How many pennies or cents are the same as seven dimes?

19. Ten counted six times equals _____ .

20. Jason did five math problems on Monday. He did ten times as many problems on Tuesday. How many problems did he do on Tuesday?

LESSON PRACTICE 5C

Find the answer by multiplying.

1. $3 \times 10 =$ _____   2. $8 \times 10 =$ _____

3. $10 \times 1 =$ _____   4. $2 \times 10 =$ _____

5. $(10)(9) =$ _____   6. $(7)(10) =$ _____

7. $10 \cdot 5 =$ _____   8. $6 \cdot 10 =$ _____

9. 10
 $\underline{\times\ 0}$

10. 10
 $\underline{\times\ 4}$

11. 10
 $\underline{\times 10}$

12. 10
 $\underline{\times\ 3}$

13. $10 \times 1 =$ _____   14. $10 \times 4 =$ _____

 $1 \times 10 =$ _____ $4 \times 10 =$ _____

15. $10 \times 2 =$ _____   16. $7 \times 10 =$ _____

 $2 \times 10 =$ _____ $10 \times 7 =$ _____

GAMMA LESSON PRACTICE 5C

Color all the boxes that have a number you would say when skip counting by 10. What kind of pattern do you see?

17.
0	1	2	3	4	5	6	7	8	9
10	11	12	13	14	15	16	17	18	19
20	21	22	23	24	25	26	27	28	29
30	31	32	33	34	35	36	37	38	39
40	41	42	43	44	45	46	47	48	49
50	51	52	53	54	55	56	57	58	59
60	61	62	63	64	65	66	67	68	69
70	71	72	73	74	75	76	77	78	79
80	81	82	83	84	85	86	87	88	89
90	91	92	93	94	95	96	97	98	99

18. How many pennies or cents are the same as five dimes?

19. Ten counted three times equals _____ .

20. The professor paid two 10-dollar bills for his new book. How much did the book cost?

SYSTEMATIC REVIEW

Find the answer by multiplying.

1. 10 · 5 = _____

2. 7 × 10 = _____

3. 10 · 2 = _____

4. (10)(10) = _____

5. 2
   × 5

6. 1 0
   × 5

7. 6
   × 2

8. 7
   × 2

9. 1
   × 3

10. 9
    × 2

11. 1 0
    × 8

12. 1 0
    × 1

13. 9 × 2 = _____

    2 × 9 = _____

14. 4 × 2 = _____

    2 × 4 = _____

15. $10 \times 3 =$ _____

3 × 10 = _____

16. $5 \times 2 =$ _____

2 × 5 = _____

QUICK REVIEW

These two-digit addition and subtraction problems can be done without regrouping. Just add or subtract the units and the tens. The first one has been done for you.

Add or subtract.

17. 2 1
    + 3 2
    ─────
     5 3

18. 4 3
    + 4 3
    ─────

19. 2 8
    − 1 6
    ─────

20. 8 9
    − 5 1
    ─────

21. Jessica slept 7 hours a day for the last 10 days. How much sleep did she get in 10 days?

22. Jessica's little sister Julie still takes naps. She got 20 more hours of sleep than Jessica did during the last 10 days. How much sleep did Julie get during that time? You will need to use your answer from #21.

SYSTEMATIC REVIEW 5E

Find the answer by multiplying.

1. 10 • 8 = _____
2. 6 × 10 = _____

3. 10 • 9 = _____
4. (10)(0) = _____

5.   5
  × 1

6.   6
  × 2

7.   8
  × 1

8.   1 0
  × 5

9.   2
  × 2

10.   2
  × 5

11. 9 × 1 = _____

    1 × 9 = _____

12. 3 × 10 = _____

    10 × 3 = _____

Rewrite using place-value notation.

13. 389 = _____ + _____ + _____
14. 72 = _____ + _____

Add or subtract.

15. 46
    +22

16. 51
    +12

17. 37
    −23

18. 94
    −43

19. How many cents are the same as eight dimes? _____

20. There are four people in our family. How many fingers do we have in all?

21. Grandma made six cherry pies and four apple pies. Aunt Mona cut each pie into 10 pieces. How many pieces of pie did she have when she was finished?

22. Noah bought nine quarts of milk. How many pints of milk does he have?

SYSTEMATIC REVIEW 5F

Find the answer by multiplying.

1. 4 • 1 = _____

2. 2 × 10 = _____

3. 10 • 3 = _____

4. (10)(9) = _____

5. 6
   × 2

6. 2
   × 8

7. 1 0
   × 7

8. 1 0
   × 1

9. 3
   × 2

10. 4
    × 2

11. 1
    × 6

12. 9
    × 0

Rewrite using place-value notation.

13. 164 = _____ + _____ + _____

14. 58 = _____ + _____

Add or subtract.

15. 52
    −20

16. 64
    +13

17. 35
    +34

18. 14
    −12

19. What is five counted 10 times?

20. Shane has nine dimes. How many cents does he have?

21. Max has 5 dollars. Wayne has 10 times as much money as Max. How many dollars does Wayne have? How much money do Max and Wayne have altogether?

22. Karyn filled eight quart jars with jam. How many pints of jam did she make?

APPLICATION AND ENRICHMENT

5G

Skip count by ten. Start at the star and connect the dots all the way to 200. Use the picture to practice skip counting by ten.

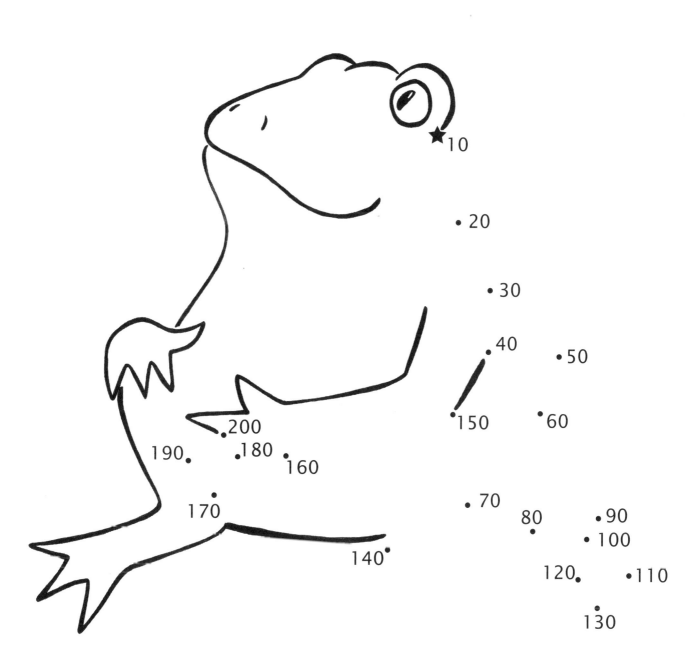

Freddy Frog ate 10 flies every Friday.

How many flies did Freddy eat in five Fridays?

Here is a pictograph for you to draw. Draw the correct number of dimes after each person's name. Line the dimes up so you can easily see who has more or fewer dimes. Here is the information you need.

Aiden - 3 dimes
Willow - 6 dimes
Connor - 3 dimes
Dani - 8 dimes
Petra - 4 dimes

1. Who has the most dimes?

2. Which two people have the same number of dimes?

3. Multiply by 10 to find how much money Willow has.

4. How many more dimes does Petra need to have the same number as Dani?

5. Multiply by 10 to find how much money Connor has.

6. Challenge: Can you use skip counting to find the total number of cents shown on the pictograph?

LESSON PRACTICE 6A

Find the answer by multiplying.

1. $5 \times 4 =$ _____
2. $5 \times 9 =$ _____

3. $5 \times 8 =$ _____
4. $5 \times 10 =$ _____

5. $(2)(5) =$ _____
6. $(5)(5) =$ _____

7. $5 \cdot 1 =$ _____
8. $5 \cdot 3 =$ _____

9. 7
 $\underline{\times\,5}$

10. 0
 $\underline{\times\,5}$

11. 6
 $\underline{\times\,5}$

12. 5
 $\underline{\times\,5}$

13. $5 \times 10 =$ _____
 $10 \times 5 =$ _____

14. $5 \times 7 =$ _____
 $7 \times 5 =$ _____

15. $5 \times 3 =$ _____
 $3 \times 5 =$ _____

16. $5 \times 6 =$ _____
 $6 \times 5 =$ _____

Color all the boxes that have a number you would say when skip counting by 5. Continue the pattern all the way to 95.

17.

0	1	2	3	4	5	6	7	8	9
10	11	12	13	14	15	16	17	18	19
20	21	22	23	24	25	26	27	28	29
30	31	32	33	34	35	36	37	38	39
40	41	42	43	44	45	46	47	48	49
50	51	52	53	54	55	56	57	58	59
60	61	62	63	64	65	66	67	68	69
70	71	72	73	74	75	76	77	78	79
80	81	82	83	84	85	86	87	88	89
90	91	92	93	94	95	96	97	98	99

18. How many pennies or cents are the same as two nickels?

19. Five counted four times equals _____ .

20. Everyone put their hands on the table. If Ted counted eight hands, how many fingers are on the table?

LESSON PRACTICE 6B

Find the answer by multiplying.

1. $5 \times 8 =$ _____

2. $5 \times 4 =$ _____

3. $5 \times 6 =$ _____

4. $5 \times 1 =$ _____

5. $(2)(5) =$ _____

6. $(9)(5) =$ _____

7. $5 \cdot 3 =$ _____

8. $7 \cdot 5 =$ _____

9. 5
   $\underline{\times 5}$

10. 5
    $\underline{\times 0}$

11. 10
    $\underline{\times 5}$

12. 5
    $\underline{\times 4}$

13. $5 \times 2 =$ _____

    $2 \times 5 =$ _____

14. $5 \times 8 =$ _____

    $8 \times 5 =$ _____

15. $9 \times 5 =$ _____

    $5 \times 9 =$ _____

16. $5 \times 1 =$ _____

    $1 \times 5 =$ _____

LESSON PRACTICE 6B

Write the factors in the parentheses and the product in the oval. Write the problem two ways.

17.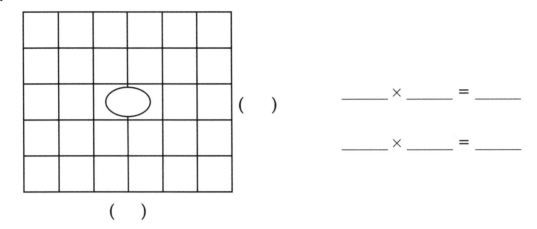

___ × ___ = ___

___ × ___ = ___

Skip count and write the missing numbers. Then fill in the missing factors under the lines.

18.
$\dfrac{0}{5\times\underline{}}$ $\dfrac{}{5\times1}$ $\dfrac{10}{5\times\underline{}}$ $\dfrac{15}{5\times\underline{}}$ $\dfrac{}{5\times4}$ $\dfrac{25}{5\times\underline{}}$

$\dfrac{}{5\times6}$ $\dfrac{}{5\times7}$ $\dfrac{40}{5\times\underline{}}$ $\dfrac{45}{5\times\underline{}}$ $\dfrac{}{5\times10}$

19. Five counted eight times equals _____ .

20. George has five nickels. Each nickel is worth 5¢. How much money does George have?

LESSON PRACTICE 6C

Find the answer by multiplying.

1. $5 \times 2 =$ _____
2. $6 \times 5 =$ _____

3. $5 \times 10 =$ _____
4. $0 \times 5 =$ _____

5. $(5)(1) =$ _____
6. $(7)(5) =$ _____

7. $5 \cdot 5 =$ _____
8. $4 \cdot 5 =$ _____

9. $\quad 5$
   $\underline{\times\, 3}$

10. $\quad 8$
    $\underline{\times\, 5}$

11. $\quad 5$
    $\underline{\times\, 9}$

12. $\quad 5$
    $\underline{\times\, 6}$

13. $5 \times 4 =$ _____

    $4 \times 5 =$ _____

14. $5 \times 10 =$ _____

    $10 \times 5 =$ _____

15. $5 \times 7 =$ _____

    $7 \times 5 =$ _____

16. $5 \times 0 =$ _____

    $0 \times 5 =$ _____

LESSON PRACTICE 6C

Color all the boxes that have a number you would say when skip counting by 5. Continue the pattern all the way to 95.

17.

0	1	2	3	4	5	6	7	8	9
10	11	12	13	14	15	16	17	18	19
20	21	22	23	24	25	26	27	28	29
30	31	32	33	34	35	36	37	38	39
40	41	42	43	44	45	46	47	48	49
50	51	52	53	54	55	56	57	58	59
60	61	62	63	64	65	66	67	68	69
70	71	72	73	74	75	76	77	78	79
80	81	82	83	84	85	86	87	88	89
90	91	92	93	94	95	96	97	98	99

18. How many pennies or cents are the same as eight nickels?

19. Five counted 10 times equals _____ .

20. Lindsay read five pages of her book every day for a week (seven days). How many pages did she read?

6D

SYSTEMATIC REVIEW

Find the answer by multiplying.

1. $5 \cdot 6 =$ _____

2. $10 \times 7 =$ _____

3. $2 \cdot 6 =$ _____

4. $(5)(8) =$ _____

5. $\begin{array}{r} 9 \\ \times\, 5 \\ \hline \end{array}$

6. $\begin{array}{r} 10 \\ \times\, 6 \\ \hline \end{array}$

7. $\begin{array}{r} 8 \\ \times\, 2 \\ \hline \end{array}$

8. $\begin{array}{r} 7 \\ \times\, 5 \\ \hline \end{array}$

9. $\begin{array}{r} 0 \\ \times\, 3 \\ \hline \end{array}$

10. $\begin{array}{r} 9 \\ \times\, 1 \\ \hline \end{array}$

11. $\begin{array}{r} 5 \\ \times\, 4 \\ \hline \end{array}$

12. $\begin{array}{r} 10 \\ \times\, 3 \\ \hline \end{array}$

13. $3 \times 5 =$ _____

    $5 \times 3 =$ _____

14. $7 \times 2 =$ _____

    $2 \times 7 =$ _____

15. $10 \times 5 =$ _____

    $5 \times 10 =$ _____

16. $1 \times 2 =$ _____

    $2 \times 1 =$ _____

GAMMA SYSTEMATIC REVIEW 6D

SYSTEMATIC REVIEW 6D

QUICK REVIEW

When adding, if the sum of the units place is more than nine, the extra 10 must be regrouped to the tens place. It can be helpful to rewrite the problem with place-value notation.

Example 1

```
      1        10
    3 4   →   30 + 4
  + 2 8   → + 20 + 8
    6 2   →   60 + 2
```

Example 2

```
      1        10
    4 5   →   40 + 5
  + 3 5   → + 30 + 5
    8 0   →   80 + 0
```

Add. Regroup when necessary.

17. 2 5
   + 3 6

18. 7 8
   + 3 4

19. 4 9
   + 5 1

20. 6 5
   + 1 5

21. At one time, stamps cost 5¢ apiece. How many letters could be mailed for 35¢? (skip count)

22. I called Amy, and we talked for 25 minutes. The next day, Amy called me, and we talked for 58 minutes. For how many minutes did Amy and I talk in all?

SYSTEMATIC REVIEW

Find the answer by multiplying.

1. $5 \cdot 5 =$ _____

2. $1 \times 5 =$ _____

3. $2 \cdot 9 =$ _____

4. $(10)(10) =$ _____

5. $\begin{array}{r} 10 \\ \times\ 8 \\ \hline \end{array}$

6. $\begin{array}{r} 5 \\ \times\ 2 \\ \hline \end{array}$

7. $\begin{array}{r} 6 \\ \times\ 5 \\ \hline \end{array}$

8. $\begin{array}{r} 9 \\ \times\ 5 \\ \hline \end{array}$

9. $\begin{array}{r} 7 \\ \times\ 1 \\ \hline \end{array}$

10. $\begin{array}{r} 2 \\ \times\ 3 \\ \hline \end{array}$

11. $\begin{array}{r} 8 \\ \times\ 2 \\ \hline \end{array}$

12. $\begin{array}{r} 9 \\ \times\ 0 \\ \hline \end{array}$

13. $4 \times 5 =$ _____

    $5 \times 4 =$ _____

14. $10 \times 2 =$ _____

    $2 \times 10 =$ _____

15. 5 × 7 = _____

    7 × 5 = _____

16. 5 × 3 = _____

    3 × 5 = _____

Add.

17. 27
    +34
    ____

18. 19
    +13
    ____

19. 61
    +29
    ____

20. 47
    +37
    ____

21. What is five counted three times?

22. Oscar hiked nine miles on the first day of the trip. By the end of the trip, he had hiked five times as far. How many miles did he hike altogether?

23. Nate conducted 18 experiments last month and 19 this month. How many experiments did he do altogether?

24. Andrea picked 15 daisies and gave 5 to her brother. She then went out and picked 15 more daisies. How many daisies does she have now?

SYSTEMATIC REVIEW

Find the answer by multiplying.

1. $5 \cdot 0 =$ _____

2. $5 \times 10 =$ _____

3. $8 \cdot 5 =$ _____

4. $(9)(10) =$ _____

5. 10
   $\underline{\times\ 4}$

6. 2
   $\underline{\times\ 6}$

7. 5
   $\underline{\times\ 2}$

8. 5
   $\underline{\times\ 3}$

9. 6
   $\underline{\times\ 0}$

10. 8
    $\underline{\times\ 1}$

11. 2
    $\underline{\times\ 4}$

12. 5
    $\underline{\times\ 7}$

SYSTEMATIC REVIEW 6F

Add or subtract. The subtraction problems do not require regrouping.

13. 61
    − 30

14. 28
    + 23

15. 49
    + 14

16. 35
    + 64

17. 57
    + 27

18. 24
    − 13

19. 88
    − 24

20. 83
    + 9

21. What is nine counted five times?

22. Kym has eight nickels. How much money does he have?

23. Trevor made eight quarts of punch for the party. How many pints did he make?

24. Forty-three lights burned out on the big Christmas tree downtown. Then 29 more lights burned out. How many lights need to be replaced?

The repair crew could find only 10 replacement bulbs. How many do they still need?

APPLICATION AND ENRICHMENT

6G

The answers to multiplication facts can be used to make interesting patterns. Follow the directions to make a pattern using the two facts.

Multiply each number in the top row by two. Write the answer in the bottom row. The first three have been done for you.

0	1	2	3	4	5	6	7	8	9	10
0	2	4								

Look at the numbers in the bottom row of the chart. Start at 0 on the circle and connect the dots in order, using the numbers in the bottom row. When you get to 10, use only the number in the units place each time. The pattern is started for you.

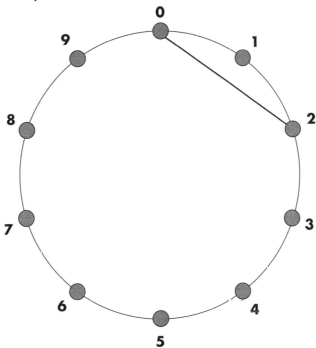

1. How many times did you go around the circle? _____

2. How many sides does the new shape have? _____

3. Challenge: What is the name of the new shape? _____

APPLICATION AND ENRICHMENT 6G

What happens if you make a pattern with the five facts? Multiply each number in the top row by five. Write the answer in the bottom row.

0	1	2	3	4	5	6	7	8	9	10
0	5									

Look at the numbers in the bottom row. Start at 0 on the circle and connect the dots in order, using the numbers in the bottom row. Keep connecting the numbers in the units place.

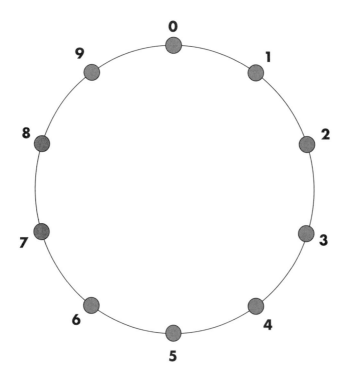

What does your pattern look like? Circle the answer.

line   square   triangle   star

LESSON PRACTICE 7A

Given the length of the sides, find the product, or area. Write it in the oval. The first one has been done for you.

1.
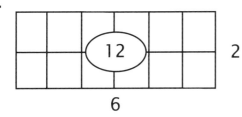

2.
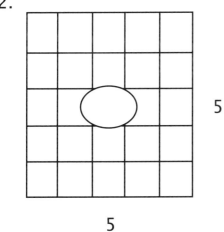

Multiply to find the product, or area. Check your answer by counting the squares inside the shape. (The drawings are not actual size.)

3.
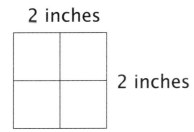

Area = _____ square inches

LESSON PRACTICE 7A

4. 2 inches / 3 inches

Area = ___ square inches

5. 5 feet / 2 feet

Area = _____ square feet

6. 1 meter / 1 meter

A meter (metric unit of length) is a little longer than a yard.

Area = _____ square meter

7. Sam wants a garden around his mailbox. If the space is five feet wide and eight feet long, what is its area?

8. A room is 10 feet long and 9 feet wide. The floor tiles are each one foot square. How many are needed to cover the floor?

LESSON PRACTICE 7B

Multiply to find the product, or area. Check your answer by counting the squares. The first one has been done for you.

1.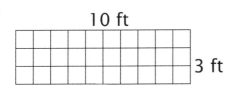

   Area = __30 sq ft__

2.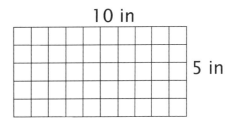

   Area = _____ sq in

3.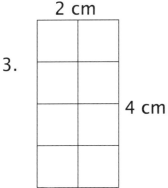

   Area = _____ sq cm

4.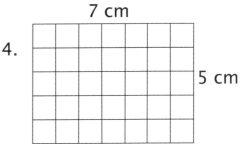

   Area = _____ sq cm

A centimeter (cm) is a metric measure that is a little less than one half of an inch.

Multiply to find the area.

5.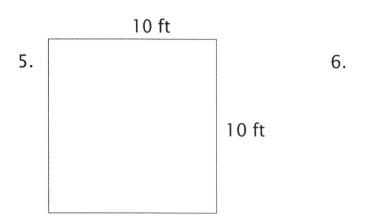

Area = _____ sq ft

6.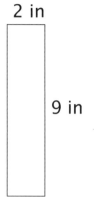

Area = _____ sq in

7. A new town will be built on a large square of land that measures five miles on each side. How many square miles will the town contain?

8. A sheet of paper is five inches wide and six inches long. What is its area?

LESSON PRACTICE 7C

Multiply to find the area. Label each answer using the abbreviations sq ft, sq in, or sq mi.

1. 6 ft / 1 ft

   Area = _____

2. 2 in / 2 in

   Area = _____

3. 10 mi (miles) / 2 mi

   Area = _____

4. 10 ft / 4 ft

   Area = _____

5.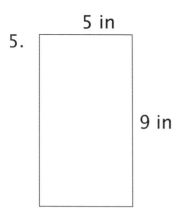

Area = _____

6.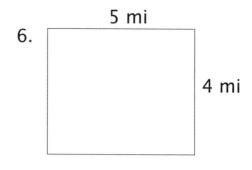

Area = _____

7. A bag of fertilizer covers 100 square feet. Will that be enough for Pam's 10-foot by 10-foot garden?

8. Thomas has a bag of square blocks. He wants to build a rectangle that has two blocks on one side and eight blocks on the other side. How many blocks will he need?

SYSTEMATIC REVIEW

7D

Find the area of each rectangle.

1. 10 in, 8 in

    Area = _____

2. 3 ft, 5 ft

    Area = _____

3. 1 mi, 1 mi

    Area = _____

Multiply.

4. 5 • 9 =

5. 10 × 6 =

6. 5 • 7 =

7. (0)(6) =

8. 1
   × 8

9. 7
   × 2

10. 2
   × 5

11. 5
   × 6

SYSTEMATIC REVIEW 7D

Add or subtract. The subtraction problems do not require regrouping.

12. 35
    + 56

13. 48
    − 24

14. 78
    + 9

15. 84
    − 13

16. How many square miles are in a park that is two miles wide and nine miles long?

17. Mom made six quarts of strawberry jam. How many pints did she make?

18. There is a patch of sunlight on the floor that measures three feet by two feet. Two kittens are curled up in each square foot of sunlight. How many kittens are sunning themselves? (This is a two-step problem.)

SYSTEMATIC REVIEW 7E

Find the area of each rectangle.

1.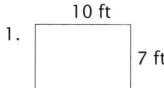

   Area = _____

2. 5 mi

   5 mi

   Area = _____

3.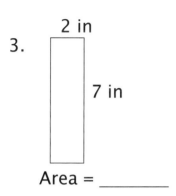

   Area = _____

Multiply.

4. 1 • 9 =

5. 10 × 4 =

6. 5 • 8 =

7. (2)(6) =

8.   10
    × 1
   ―――

9.   5
    × 3
   ―――

10.   10
     × 7
    ―――

11.   9
     × 2
    ―――

Add or subtract. The subtraction problems do not require regrouping.

12. 79
    − 21

13. 32
    + 59

14. 63
    − 30

15. 45
    + 45

16. Logan has six library books that are overdue. The fine is 5¢ apiece. How much is owed in all?

17. The area of the smaller carpet is five square feet. The area of the larger carpet is three times as large. What is the area of the larger carpet?

18. Aiden collected 16 acorns, and Willow collected 26 acorns. How many acorns were added to their collection?

    Twelve acorns were lost. How many are left?

SYSTEMATIC REVIEW

7F

Find the area of each rectangle. If the units in an area problem are not given, label the answer as square units.

1. 4 in × 1 in

   Area = _____

2. 10 mi × 10 mi

   Area = _____

3. 5 × 9

   Area = _____ square units

Multiply.

4. 10 • 3 = _____

5. 5 × 2 = _____

6. 4 • 5 = _____

7. (10)(2) = _____

8. 1
   × 1

9. 8
   × 2

10. 2
    × 4

11. 5
    × 7

SYSTEMATIC REVIEW 7F

Add or subtract. The subtraction problems do not require regrouping.

12. 67
    + 53

13. 19
    + 12

14. 93
    + 8

15. 61
    − 40

16. Jen has two cans of paint. Each can holds enough paint to cover 50 square feet. Her bathroom ceiling is a square that measures 10 feet on each side. Does she have enough paint to paint the ceiling?

17. Forty-five ants went marching down into the ground to get out of the rain. If there were five ants in each row, how many rows of ants were marching? (skip count)

18. Val had 13 pints of grape jelly and 12 pints of raspberry jelly. She used five pints of jelly to make sandwiches for a party. How may pints of jelly does she have left over?

Use skip counting if needed to find how many *quarts* of jelly are left.

APPLICATION AND ENRICHMENT 7G

Find the area of each garden. If you wish, you may draw or color the plants that you think should be planted in each garden.

1.

2 ft

10 ft

Area = _____

2.

4 ft

5 ft

Area = _____

3.

1 ft

20 ft

Area = _____

4. Can rectangles with different dimensions have the same area?

APPLICATION AND ENRICHMENT 7G

Here is a pictograph for you to draw. Write the name of the items in the rectangles at the left of the graph. Draw the correct number of nickels after each item. Line the nickels up so you can easily see which item costs the most or the least. Here is the information you need.

pencil - 3 nickels
apple - 6 nickels
toy car - 5 nickels
crayons - 8 nickels
ring - 10 nickels

1. Which item costs the most?

2. How many cents are needed to pay for the crayons?

3. Which item costs the least?

4. What is the difference between the price of the car and the price of the pencil?

5. Challenge: Use skip counting to find how many cents are equal to all the nickels shown on the pictograph.

LESSON PRACTICE

Solve for the unknown with multiplication. The first one has been done for you.

1. Two times what is equal to ten?

   $2 \times 5 = 10$

2. Five times what is equal to twenty-five?

   $5 \times \underline{} = 25$

3. Ten times what is equal to eighty?

   $10 \times \underline{} = 80$

4. One times what is equal to seven?

   $1 \times \underline{} = 7$

5. $4F = 20$ 6. $10A = 50$

7. $2K = 18$ 8. $5Y = 15$

9. $3X = 3$ 10. $2Y = 4$

LESSON PRACTICE 8A

11. 1R = 10

12. 8W = 16

13. 7H = 14

14. 5B = 30

15. 10X = 100

16. 5G = 45

17. There are 12 ears in the room. How many people are in the room?

18. Gumdrops are two cents apiece. How many gumdrops can Jason buy for 18 cents?

19. Each vase has 10 roses. There are 40 roses. How many vases are there?

20. Tyler earns $5 a week doing chores. How many weeks will it take him to earn $50?

LESSON PRACTICE 8B

Solve for the unknown with multiplication.

1. Two times what is equal to twenty?

    2 × ____ = 20

2. Five times what is equal to thirty-five?

    5 × ____ = 35

3. Ten times what is equal to ninety?

    10 × ____ = 90

4. Six times what is equal to zero?

    6 × ____ = 0

5. $4F = 8$

6. $7A = 70$

7. $5K - 30$

8. $8Y = 8$

9. $2X = 12$

10. $10Y = 60$

GAMMA LESSON PRACTICE 8B

LESSON PRACTICE 8B

11. 8R = 40

12. 10W = 20

13. 3H = 6

14. 1B = 4

15. 2X = 20

16. 8G = 0

17. Emily has 14 hair clips in her drawer. She wears two of them every day. If she wears a different pair each day, how many days will it take to wear all of her hair clips?

18. Everyone put their hands on the table. Morgan counted 16 hands. How many people were there?

19. The pet shop owner plans to put five canaries in each cage. He has 35 canaries. How many cages does he need?

20. Erin has $100 to spend on Christmas gifts. She needs to buy 10 gifts. How much can she spend on each gift?

LESSON PRACTICE

Solve for the unknown with multiplication.

1. Two times what is equal to twelve?

    2 × _____ = 12

2. Five times what is equal to ten?

    5 × _____ = 10

3. Ten times what is equal to forty?

    10 × _____ = 40

4. One times what is equal to six?

    1 × _____ = 6

5. 5F = 25   6. 8A = 80

7. 9K = 9   8. 5Y = 45

9. 2X = 16   10. 3Y = 15

LESSON PRACTICE 8C

11. 10R = 100

12. 9W = 18

13. 3H = 30

14. 5B = 40

15. 5X = 0

16. 2G = 2

17. There are 10 mittens on the shelf. How many children can have warm hands?

18. Twenty students are going on a field trip. Five can ride in each car. How many cars are needed?

19. Mom bought eight pints of strawberries. How many quarts does she have?

20. The grocery bill came to $90. How many 10-dollar bills would be needed to pay for the groceries?

SYSTEMATIC REVIEW 8D

Solve for the unknown.

1. $2X = 20$
2. $10T = 60$
3. $5A = 35$
4. $5D = 5$
5. $7F = 14$
6. $6B = 30$

Find the area of each rectangle.

7. 5 in × 1 in

   A = _____

8. 5 ft × 5 ft

   A = _____

9. 5 mi × 4 mi

   A = _____

Add.

10.   6 1
    + 2 9

11.   7 2
    + 3 8

SYSTEMATIC REVIEW 8D

12.   4 4
    + 5 5

13.   8 6
    + 7 3

QUICK REVIEW

When subtracting, it is sometimes necessary to regroup from the tens place. Study the following example.

Example 1

$$\begin{array}{ccccc} 42 & 40+2 & 3\overset{10}{\cancel{0}}+2 & 30+12 \\ -18 & -(10+8) & -(10+8) & -(10+8) \\ \hline 24 & & & 20+4 \end{array}$$

Subtract, regrouping as necessary. The first one has been done for you.

14. $^2\cancel{3}\,^14$
   − 1 6
     1 8

15.   7 4
    − 3 8

16.   7 1
    − 5 9

17.   6 7
    − 2 5

18. Sam's mom made 65 cookies for the bake sale. If 37 of the cookies were sold, how many are left?

SYSTEMATIC REVIEW

Solve for the unknown.

1. $5X = 40$

2. $10T = 70$

3. $2A = 18$

4. $1D = 1$

5. $3F = 6$

6. $9B = 45$

Find the area of each rectangle.

7. 10 in, 4 in

    A = _____

8. 2 ft, 2 ft

    A = _____

9. 10 ft, 8 ft

    A = _____

SYSTEMATIC REVIEW 8E

Add.

10. 35
    +65

11. 53
    +61

12. 99
    +22

13. 26
    +48

Subtract.

14. 17
    − 9

15. 45
    −26

16. 93
    −34

17. 52
    −42

18. Mary canned two pints of tomatoes. Martha canned five times as many pints. How many pints did Martha can?

19. Al had 12 peanuts. He shared them equally with his friend. How many peanuts did each person get?

20. The state park measures 10 miles on each side. What is the area of the state park?

SYSTEMATIC REVIEW

Solve for the unknown.

1. $4X = 20$
2. $2T = 4$
3. $10A = 20$
4. $5D = 25$
5. $9F = 0$
6. $3B = 30$

Find the area of each rectangle.

7. 4 mi / 2 mi

   A = _____

8. 1 in / 1 in

   A = _____

9. 5 ft / 6 ft

   A = _____

SYSTEMATIC REVIEW 8F

Add.

10.   21
    + 58

11.   15
    + 17

12.   84
    + 46

13.   76
    + 75

Subtract.

14.   24
    − 15

15.   53
    − 35

16.   65
    − 19

17.   37
    − 8

18. Joseph received two boxes of blocks for his birthday. The first box had 68 pieces, and the second had 94 pieces. How many blocks did he receive?

19. Steve paid his three older boys $15 for their work. If each boy received the same amount, how much did each get?

20. Hannah Joy picked two pints of blueberries for each member of her family, including herself. She has two brothers, one sister, and her mom and dad. How many pints did she pick?

APPLICATION AND ENRICHMENT

8G

Help to write the word problems. Fill in the blank and solve for the unknown to find the answer. Write your answer in the box.

1. Justin collected _____ . He put two of them in each box. There are 12 items in Justin's collection. How many boxes does he need?

2. Tim watched 15 _____ fly into his room through the windows. Five of them came in each window. How many windows are in Tim's room?

3. Rena made 18 _____ . She gave them to Callie and Rachel. She gave the same number to each girl. How many did each girl receive?

4. Julia has $80. She wants to buy several _____ . They cost $10 apiece. How many can she buy?

APPLICATION AND ENRICHMENT 8G

You can use charts called *Carroll diagrams* to figure out the answers to word puzzles. These kinds of thinking problems are called *logic problems.* There are hints under the first chart to get you started.

1. Lucky, Mia, and Prince are each a different kind of pet. Lucky does not have fins. Mia cannot swim. Tell what kind of animal each one is.

	fish	dog	canary
Lucky	x		
Mia	x	x	
Prince	o		

Lucky does not have fins. Put an X by Lucky's name under fish. Mia can't swim. Since fish and dogs can swim, put X's by Mia's name to show that she is not a fish or a dog.

Look at your chart. The fish has to be Prince, so draw a circle in that column. You can put Xs in the other two columns by Prince's name if you wish.

The only space left for Mia is canary. Then the only pet left is Lucky, and the only choice left is dog.

2. Hope, Alana, and Jared all have different hobbies. Hope and Alana are not good at catching things. Jared and Alana don't like art. What is each person's hobby?

	drawing	juggling	piano
Hope			
Alana			
Jared			

LESSON PRACTICE

Skip count by 9. Write the correct numbers in the squares with lines.

1.

									9
									__
									__
									__
									45
									__
									__
									__
									81
									__

Skip count and write the numbers.

2. 9, ____, ____, ____, ____, 54, ____, ____, ____, 90

Find the missing multiples of 2 and 9 in the equivalent fractions.

3. $\dfrac{2}{9} = \dfrac{4}{\_\_} = \dfrac{\_\_}{27} = \dfrac{8}{\_\_} = \dfrac{\_\_}{45} = \dfrac{\_\_}{54} = \dfrac{14}{\_\_} = \dfrac{\_\_}{72} = \dfrac{\_\_}{81} = \dfrac{20}{\_\_}$

LESSON PRACTICE 9A

Use skip counting to solve the word problems.

4. There were six baseball teams in the league. Each team had nine players. How many players were in the league?

5. I gave a dollar to the cashier, and she gave me an apple and nine dimes. How many cents worth of change did I get?

6. Nine children each ate three slices of pizza. How many slices of pizza were eaten?

LESSON PRACTICE 9B

Skip count by 9. Write the correct numbers in the squares with lines.

1.

									___
									18
									___
									___
									54
									___
									___
									___

Skip count and write the numbers.

2. ___, ___, 27, ___, ___, ___, ___, ___, ___, ___

Find the missing multiples of 5 and 9 in the equivalent fractions.

3. $\dfrac{5}{9} = \dfrac{}{18} = \dfrac{15}{} = \dfrac{20}{} = \dfrac{}{45} = \dfrac{}{63} = \dfrac{40}{} = \dfrac{}{} = \dfrac{}{90}$

LESSON PRACTICE 9B

Use skip counting to solve the word problems.

4. The school field day had nine races. Four children ran in each race. How many children raced in all?

5. Jonathan made labels for his rock collection. There are nine labels on a sheet, and he used eight sheets of labels. How many rocks did Jonathan label?

6. Mr. Aldrich needs carpet for his new office. The floor measures 9 feet by 10 feet. What is the area of his office?

LESSON PRACTICE

Skip count by 9. Write the correct numbers in the squares with lines.

1.

									___
									___
									___
									36
									___
									___
									___
									___
									___
									___

Skip count and write the numbers.

2. ___, 18, ___, ___, ___, ___, ___, ___, ___, ___

Find the missing multiples of 9 and 10 in the equivalent fractions.

3. $\dfrac{9}{10} = \dfrac{18}{} = \dfrac{}{30} = \dfrac{}{40} = \dfrac{}{} = \dfrac{54}{} = \dfrac{}{} = \dfrac{72}{} = \dfrac{}{} = \dfrac{}{100}$

LESSON PRACTICE 9C

Use skip counting to solve the word problems.

4. We are tiling our kitchen floor. The floor needs nine tiles across and nine tiles up. How many tiles should we buy?

5. Ghengis Khan captured nine cities a year. How many cities did he capture in five years?

6. Mary wrote nine letters a day for a week (seven days). How many letters did she write?

SYSTEMATIC REVIEW

Skip count and write the numbers.

1. 9, ____, ____, ____, ____, ____, ____, ____, ____, ____

2. ____, ____, 15, ____, ____, ____, ____, ____, ____, ____

Solve for the unknown.

3. 5D = 25

4. 6F = 60

5. 2B = 18

Multiply.

6. (1)(7) = ____

7. 5 × 4 = ____

8. 10 • 8 = ____

Find the area of each rectangle.

9. [7 in by 2 in rectangle]
A = ____

10. [10 mi by 10 mi square]
A = ____

SYSTEMATIC REVIEW 9D

11. 5 in, 3 in

A = _____

Add or subtract.

12. 25
    +62

13. 91
    −45

14. 16
    +17

15. 86
    −73

16. I gave James three nickels. How much money did I give him?

17. Rachel has four nickels. How many cents does she have?

18. Balloons cost 10¢ apiece. How much will it cost to give each of my eight friends a balloon?

19. Jeff bought 14 arrows for his bow. He got 29 more arrows for his birthday. How many arrows does he have now?

20. There are nine baseball players on a team. How many players are needed for two teams?

SYSTEMATIC REVIEW

9E

Skip count and write the numbers.

1. ____, 18, ____, ____, ____, ____, ____, ____, ____, ____

Find the missing multiples of 5 and 10 in the equivalent fractions.

2. $\dfrac{5}{10} = \dfrac{}{20} = \dfrac{15}{40} = \dfrac{}{60} = \dfrac{}{} = \dfrac{}{} = \dfrac{40}{} = \dfrac{}{} = \dfrac{}{100}$

Solve for the unknown.

3. $3X = 30$

4. $4A = 8$

5. $5Q = 35$

Multiply.

6. $(0)(8) = $ ____

7. $9 \times 10 = $ ____

8. $6 \cdot 5 = $ ____

Find the area of each rectangle.

9. 10 ft, 1 ft

   A = ____

10. 5 in, 5 in

    A = ____

GAMMA SYSTEMATIC REVIEW 9E

SYSTEMATIC REVIEW 9E

11.

5 in
2 in

A = _____

Add or subtract.

12. 13
    + 47

13. 75
    − 18

14. 43
    + 59

15. 94
    − 55

16. Nine people can ride in each van. How many vans are needed for 54 people? (skip count)

17. Joseph bought seven pencils. Each one cost a dime. How much money did he spend?

18. Rings cost 5¢ each. How many can you buy for 45¢?

19. Terry had 44 tons of silage in one silo and 26 tons in his other silo. In one month, he fed 12 tons of silage to his cows. How many tons of silage did he have left?

20. Mr. Gates raised bees. They made six quarts of honey. How many pint jars can he fill?

SYSTEMATIC REVIEW

9F

Skip count and write the numbers.

1. 9, ___, ___, ___, ___, ___, ___, ___, ___, ___

Skip count and write the missing numbers. Then fill in the missing factors under the lines.

2.

$\dfrac{0}{(2)(0)}$ $\dfrac{2}{(2)(\_\_)}$ $\dfrac{\_\_}{(2)(\_\_)}$ $\dfrac{\_\_}{(2)(3)}$ $\dfrac{8}{(2)(\_\_)}$ $\dfrac{\_\_}{(2)(5)}$

$\dfrac{\_\_}{(2)(\_\_)}$ $\dfrac{\_\_}{(2)(7)}$ $\dfrac{16}{(2)(\_\_)}$ $\dfrac{\_\_}{(2)(9)}$ $\dfrac{\_\_}{(2)(10)}$

Solve for the unknown.

3. 8Y = 40

4. 9E = 90

5. 1R = 4

Multiply.

6. (9)(1) = _____

7. 2 × 10 = _____

8. 2 • 9 = _____

SYSTEMATIC REVIEW 9F

Find the area of each rectangle.

9. 3 mi / 2 mi

A = ____

10. 2 ft / 2 ft

A = ____

11. 9 in / 5 in

A = ____

Add or subtract.

12. 27
    − 19

13. 61
    + 39

14. 54
    + 47

15. 82
    − 73

16. Scott needs to earn $81. He can earn $9 a day. How many days must he work to earn $81?

17. Aaron found six nickels in his pocket. How much money does he have?

18. Aaron's brother found six dimes under the bed. How much money did he find? How much money did Aaron (#17) and his brother find altogether?

19. Jack fetched 11 pails of water from the top of the hill and 9 from the well. Jill fetched 6 pails of water from the kitchen sink. How many more pails of water did Jack fetch?

20. Ellen read a book a day for 10 weeks. How many books did she read in all?

APPLICATION AND ENRICHMENT

9G

Skip count by nine. Start at the star and connect the dots all the way to 99. Use the picture to practice skip counting by nine.

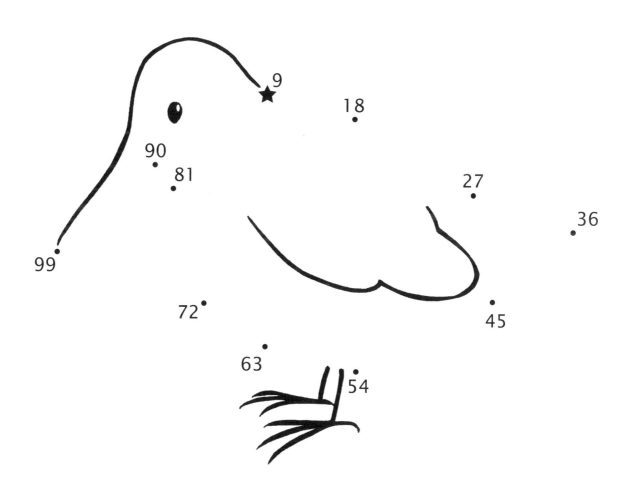

Bertha Bird has four baby birds.

She brought each baby bird nine bugs.

Skip count by nine to find how many bugs Bertha brought.

_____

APPLICATION AND ENRICHMENT 9G

Color the picture. Complete each step in the order given for best results. If you have already colored a number, do not color it again in the next step.

If you say the number when you skip count by 10 to 100, color the space yellow.
If you say the number when you skip count by 9 to 90, color the space green.
If you say the number when you skip count by 5 to 50, color the space pink.
If you say the number when you skip count by 2 to 20, color the space brown.

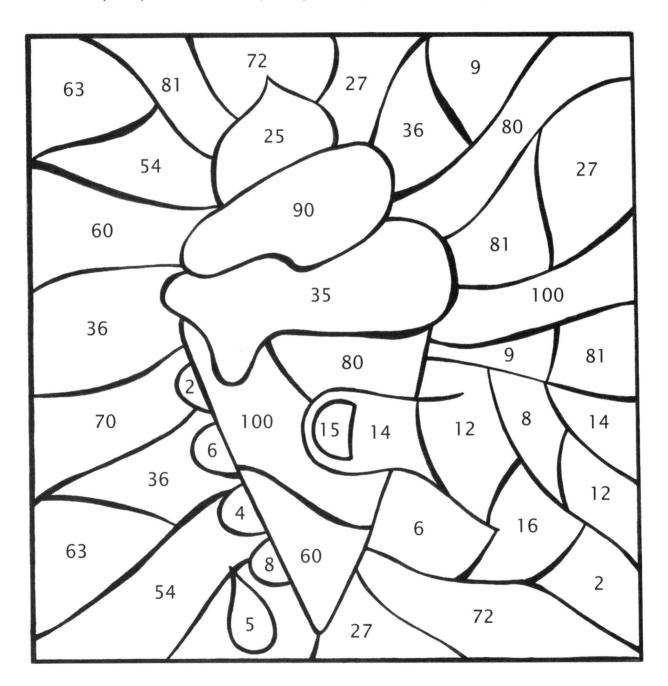

LESSON PRACTICE

10A

Find the answer by multiplying.

1. $9 \times 4 =$ _____
2. $9 \times 9 =$ _____

3. $9 \times 8 =$ _____
4. $10 \times 9 =$ _____

5. $(2)(9) =$ _____
6. $(5)(9) =$ _____

7. $9 \cdot 1 =$ _____
8. $9 \cdot 3 =$ _____

9. 9
   $\underline{\times 7}$

10. 9
    $\underline{\times 9}$

11. 6
    $\underline{\times 9}$

12. 3
    $\underline{\times 9}$

13. $9 \times 8 =$ _____
14. $9 \times 4 =$ _____

    $8 \times 9 =$ _____
    $4 \times 9 =$ _____

15. $9 \times 5 =$ _____
16. $9 \times 7 =$ _____

    $5 \times 9 =$ _____
    $7 \times 9 =$ _____

GAMMA LESSON PRACTICE 10A

LESSON PRACTICE 10A

17. Nine counted zero times equals _____ .

18. Nine counted 10 times equals _____ .

19. Nine counted two times equals _____ .

20. Nine counted six times equals _____

Color all the boxes that have a number you would say when skip counting by 9. Notice the pattern.

21.

0	1	2	3	4	5	6	7	8	9
10	11	12	13	14	15	16	17	18	19
20	21	22	23	24	25	26	27	28	29
30	31	32	33	34	35	36	37	38	39
40	41	42	43	44	45	46	47	48	49
50	51	52	53	54	55	56	57	58	59
60	61	62	63	64	65	66	67	68	69
70	71	72	73	74	75	76	77	78	79
80	81	82	83	84	85	86	87	88	89
90	91	92	93	94	95	96	97	98	99

22. Four people are ready to do the job, but nine times as many are needed. How many people are needed for the job?

LESSON PRACTICE 10B

Find the answer by multiplying.

1. $9 \times 10 =$ _____
2. $9 \times 6 =$ _____

3. $9 \times 2 =$ _____
4. $0 \times 9 =$ _____

5. $(7)(9) =$ _____
6. $(3)(9) =$ _____

7. $9 \cdot 5 =$ _____
8. $9 \cdot 8 =$ _____

9. 4
   $\underline{\times\, 9}$

10. 1
    $\underline{\times\, 9}$

11. 9
    $\underline{\times\, 9}$

12. 6
    $\underline{\times\, 9}$

13. $9 \times 3 =$ _____

    $3 \times 9 =$ _____

14. $9 \times 2 =$ _____

    $2 \times 9 =$ _____

15. $9 \times 6 =$ _____

    $6 \times 9 =$ _____

16. $9 \times 10 =$ _____

    $10 \times 9 =$ _____

LESSON PRACTICE 10B

17. Nine counted seven times equals _____ .

18. Nine counted five times equals _____ .

19. Nine counted eight times equals _____ .

20. Nine counted four times equals _____ .

Skip count and write the missing numbers. Then fill in the missing factors under the lines.

21.

$\dfrac{}{9 \cdot 0} \quad \dfrac{9}{9 \cdot \underline{}} \quad \dfrac{}{9 \cdot 2} \quad \dfrac{27}{9 \cdot \underline{}} \quad \dfrac{}{9 \cdot 4} \quad \dfrac{}{9 \cdot \underline{}}$

$\dfrac{}{9 \cdot 6} \quad \dfrac{63}{9 \cdot \underline{}} \quad \dfrac{}{9 \cdot 8} \quad \dfrac{81}{9 \cdot \underline{}} \quad \dfrac{}{9 \cdot 10}$

22. The spacecraft discovered five solar systems with nine planets each. How many planets were found altogether?

23. A quilt has nine rows with nine squares in a row. How many squares are in the quilt?

24. Samuel Stolzfus farmed nine acres of corn, nine acres of oats, and nine acres of barley. How many acres did he farm altogether?

LESSON PRACTICE

10C

Find the answer by multiplying.

1. $9 \times 1 =$ _____ 2. $9 \times 8 =$ _____

3. $9 \times 6 =$ _____ 4. $10 \times 2 =$ _____

5. $(9)(9) =$ _____ 6. $(3)(9) =$ _____

7. $5 \cdot 9 =$ _____ 8. $9 \cdot 7 =$ _____

9. 9 10. 4
   $\underline{\times 0}$ $\underline{\times 9}$

11. 10 12. 9
    $\underline{\times 9}$ $\underline{\times 8}$

13. $9 \times 1 =$ _____ 14. $9 \times 7 =$ _____

    $1 \times 9 =$ _____ $7 \times 9 =$ _____

15. $9 \times 4 =$ _____ 16. $9 \times 5 =$ _____

    $4 \times 9 =$ _____ $5 \times 9 =$ _____

LESSON PRACTICE 10C

17. Nine counted three times equals _____ .

18. Nine counted nine times equals _____ .

19. Nine counted 10 times equals _____ .

20. Nine counted zero times equals _____ .

Color all the boxes that have a number you would say when skip counting by 9. Use the chart to find 9×11.

21.

0	1	2	3	4	5	6	7	8	9
10	11	12	13	14	15	16	17	18	19
20	21	22	23	24	25	26	27	28	29
30	31	32	33	34	35	36	37	38	39
40	41	42	43	44	45	46	47	48	49
50	51	52	53	54	55	56	57	58	59
60	61	62	63	64	65	66	67	68	69
70	71	72	73	74	75	76	77	78	79
80	81	82	83	84	85	86	87	88	89
90	91	92	93	94	95	96	97	98	99

22. At one time, postage stamps cost 6¢ apiece. What did it cost to mail nine letters?

SYSTEMATIC REVIEW

Find the answer by multiplying.

1. $9 \times 5 =$ _____

2. $(7)(9) =$ _____

3. $5 \cdot 6 =$ _____

4. $8 \times 2 =$ _____

5. 10
   $\underline{\times 9}$

6. 8
   $\underline{\times 5}$

7. 3
   $\underline{\times 2}$

8. 7
   $\underline{\times 5}$

Solve for the unknown.

9. $9X = 54$ _____

10. $2A = 14$ _____

11. $10Y = 50$ _____

12. $6Q = 0$ _____

Find the area of each rectangle.

13.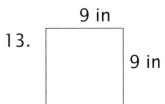

    A = _____

14.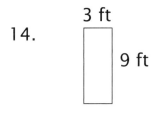

    A = _____

15.
    4 mi
    [ ] 5 mi

    A = _____

16. Michelle wants to give gifts to eight of her friends. She has $72. How much can she spend on each gift?

17. Briana took her car to the garage to have the oil changed and all four tires replaced. Each tire took 9 minutes to replace. The oil change took 15 minutes. How many minutes did the whole job take?

18. Oscar found five nickels and three dimes in his pocket. How much money did Oscar have?

19. Jason took a test of math facts. There were 35 questions on the test. He got 28 questions correct. How many questions had a wrong answer?

20. A paper has an area of 53 square inches. A piece that measures 2 inches by 6 inches is cut from the paper. Find the area of the piece that was cut out. Then subtract to find the area of the piece that is left.

10E

SYSTEMATIC REVIEW

Find the answer by multiplying.

1. $2 \times 9 =$ _____
2. $(5)(5) =$ _____

3. $7 \cdot 10 =$ _____
4. $6 \times 9 =$ _____

5. 1
   $\underline{\times 9}$

6. 9
   $\underline{\times 9}$

7. 5
   $\underline{\times 3}$

8. 9
   $\underline{\times 5}$

Solve for the unknown.

9. $2X = 20$ _____
10. $7A = 63$ _____

11. $2Y = 10$ _____
12. $3Q = 27$ _____

Find the area of each rectangle.

13.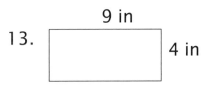

A = _____

14.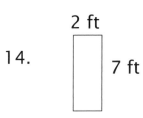

A = _____

15.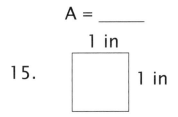

A = _____

Skip count and write the missing numbers. Then fill in the missing factors under the lines.

16.

$\dfrac{0}{9 \times \_\_}$ $\dfrac{\_\_\_}{9 \times 1}$ $\dfrac{\_\_\_}{9 \times 2}$ $\dfrac{\_\_\_}{9 \times 3}$ $\dfrac{36}{9 \times \_\_}$ $\dfrac{\_\_\_}{9 \times 5}$

$\dfrac{\_\_\_}{9 \times \_\_}$ $\dfrac{\_\_\_}{9 \times \_\_}$ $\dfrac{72}{9 \times \_\_}$ $\dfrac{\_\_\_}{9 \times 9}$ $\dfrac{\_\_\_}{9 \times 10}$

17. Olive has six dimes and seven nickels. How much money does she have?

18. What is the area of a piece of land that measures nine miles by eight miles?

19. Valerie had 13 pints of blackberry jam and 12 pints of raspberry jam. She used 7 pints of jam to make sandwiches. How many pints did she have left?

20. Robin drove 45 miles one day and 39 miles the next day. How many miles did she drive in two days?

SYSTEMATIC REVIEW

10F

Find the answer by multiplying.

1. $2 \times 4 =$ _____

2. $(5)(6) =$ _____

3. $9 \cdot 5 =$ _____

4. $3 \times 9 =$ _____

5. 8
   $\underline{\times\ 9}$

6. 9
   $\underline{\times\ 6}$

7. 10
   $\underline{\times\ 3}$

8. 4
   $\underline{\times\ 5}$

Solve for the unknown.

9. $4X = 36$

10. $2A = 18$

11. $8Y = 80$

12. $8Q = 40$

Find the area of each rectangle.

13. 9 in × 7 in

    A = _____

14. 10 ft × 10 ft

    A = _____

15.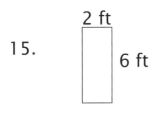

A = _____

16. Jim has seven nickels, and Lisa has four dimes. Who has more money? How much money do they have altogether?

17. Nine golfers walked out onto the golf course. Each one had nine golf clubs. During the game, three clubs were lost in a pond. How many clubs are left?

18. A children's choir has 17 boys and 25 girls. How many children are in the choir?

19. Sloths are very slow. It took the first sloth 45 minutes to cross the road. The second sloth was feeling tired and took 53 minutes to get across. How much longer did it take the second sloth to cross the road?

20. The floor tiles are one foot square. How many are needed to cover a floor that measures five feet by nine feet?

APPLICATION AND ENRICHMENT

10G

Multiply each number in the top row by nine. Write the answers in the bottom row.

0	1	2	3	4	5	6	7	8	9	10
0	9									

Look at the numbers in the bottom row. Start at 0 on the circle and connect the dots in order. Use the numbers in the units place of each answer in the bottom row.

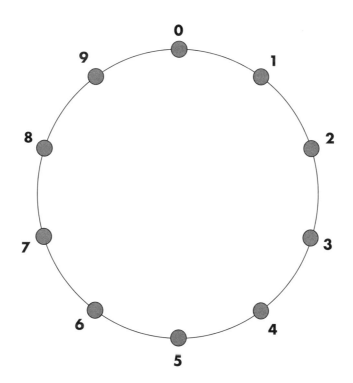

1. How many sides does the shape have?

2. Did you go around the circle like a clock (clockwise) or the opposite way (counterclockwise)? Circle the right answer.

   clockwise   counterclockwise

APPLICATION AND ENRICHMENT 10G

A *Venn diagram* is a chart made from overlapping shapes or regions. It shows you how numbers or things can belong in two different groups at the same time.

1. In region A, write the numbers that you say when you skip count by 5 from 5 to 50.

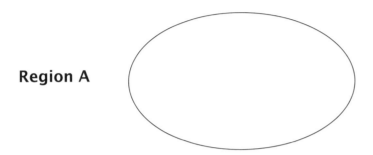

Region A

2. In region B, write the numbers that you say when you skip count by 10 from 10 to 100.

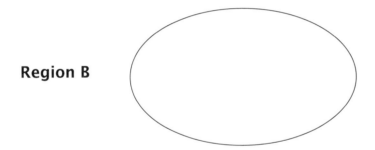

Region B

3. Now put the shapes together. The numbers that were in both go in the middle where A and B overlap. The other numbers go in the outside part of the regions where they belong. For example, 5 goes in region A, and 10 goes in the middle.

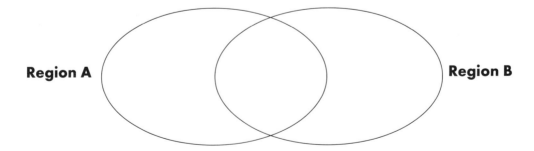

4. How many numbers are in the overlapping part of the shapes?

LESSON PRACTICE 11A

Skip count by 3. Write the correct numbers in the squares with lines.

1.

		3
		___
		___
		___
		15
		___
		___
		___
		27
		___

2.

		___
		6
		___
		___
		___
		___
		___
		24
		___
		___

Skip count and write the numbers.

3. 3, ____, ____, ____, ____, 18, ____, ____, ____, 30

Find the missing multiples of 2 and 3 in the equivalent fractions.

4. $\frac{2}{3} = \frac{___}{___} = \frac{___}{9} = \frac{8}{___} = \frac{___}{15} = \frac{___}{18} = \frac{14}{___} = \frac{___}{24} = \frac{___}{27} = \frac{20}{___}$

LESSON PRACTICE 11A

Use skip counting to solve the word problems.

5. The store has five bright red tricycles for sale. How many wheels do the tricycles have in all?

6. King Edward found that he could build a new castle every three years. How many years would it take him to build six new castles?

7. How many sides are on eight triangles?

8. How many dots are on these four squares?

LESSON PRACTICE 11B

Skip count by 3. Write the correct numbers in the squares with lines.

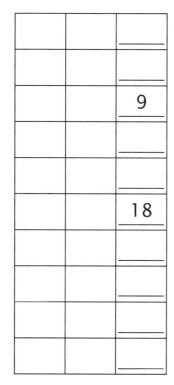

1.

2.

Skip count and write the numbers.

3. ____, 6, ____, 12, ____, ____, ____, ____, ____, ____

Find the missing multiples of 3 and 5 in the equivalent fractions.

4. $\dfrac{3}{5} = \dfrac{}{15} = \dfrac{12}{} = \dfrac{}{25} = \dfrac{}{30} = \dfrac{21}{} = \dfrac{}{40} = \dfrac{}{45} = \dfrac{30}{}$

LESSON PRACTICE 11B

Use skip counting to solve the word problems.

5. A triangle has three sides. How many sides are on three triangles?

6. One giant step is the same as three baby steps in the game "Mother, May I." If I walk nine giant steps, how many baby steps have I walked?

7. Three people rode in each car. There were eight cars. How many people went on the trip?

8. When Mom asked who wanted ice cream, all the children raised both hands. There were three children. How many hands were in the air?

LESSON PRACTICE

Skip count by 3. Write the correct numbers in the squares with lines.

1.

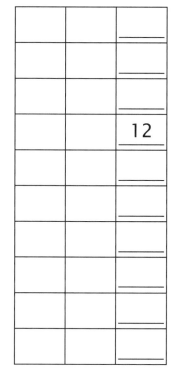

		___
		___
		___
		12
		___
		___
		___
		___
		___
		___

2.

		3
		___
		___
		___
		___
		___
		21
		___
		___
		___

Skip count and write the numbers.

3. ____, ____, ____, ____, ____, ____, ____, ____, ____, 30

Find the missing multiples of 3 and 10 in the equivalent fractions.

4. $\frac{3}{10} = \frac{\quad}{\quad} = \frac{9}{\quad} = \frac{\quad}{40} = \frac{\quad}{\quad} = \frac{18}{\quad} = \frac{\quad}{70} = \frac{\quad}{\quad} = \frac{27}{\quad} = \frac{\quad}{100}$

LESSON PRACTICE 11C

5. Our family has four children. We went shopping and bought three new shirts for each child. How many new shirts did we buy on our shopping trip?

6. Michael read three books a week for seven weeks. How many books did he read?

7. Gabriel bought Christmas gifts for his 10 friends. The gifts cost $3 each. How much did Gabriel spend?

8. Ian learned three new things yesterday and three more new things today. How many new things did he learn in all?

SYSTEMATIC REVIEW 11D

Skip count and write the numbers.

1. ___, ___, 9, ___, ___, ___, ___, ___, 27, ___

Skip count and write the missing numbers. Then fill in the missing factors under the lines

2.

$\overline{0}$ $\overline{}$ $\overline{}$ $\overline{27}$ $\overline{}$ $\overline{}$
(9)(__) (9)(1) (9)(__) (9)(__) (9)(__) (9)(5)

$\overline{}$ $\overline{}$ $\overline{}$ $\overline{}$ $\overline{90}$
(9)(__) (9)(7) (9)(__) (9)(9) (9)(__)

Solve for the unknown.

3. 9D = 27

4. 5F = 20

5. 2F = 0

6. 10B = 50

Multiply.

7. (2)(8) = _____

8. 1 × 7 = _____

9. 7 × 10 = _____

10. 9 • 4 = _____

Find the area of each rectangle.

11. 9 in by 6 in

A = _____

12. 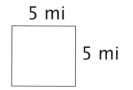 5 mi by 5 mi

A = _____

13. 7 in by 5 in

A = _____

SYSTEMATIC REVIEW 11D

QUICK REVIEW

When adding columns of numbers, it is helpful to look for combinations that make 10. Regroup in the same way as when adding two numbers. Study the examples.

Examples

```
     8 ⎫            2
     2 ⎭10         2 5 ⎫           1
     4 ⎫           1 3 ⎭10         1 4 ⎫
   + 6 ⎭10         1 5 ⎫           4 6 ⎭10
   ─────          +3 7 ⎭10        +2 4
    2 0            ─────          ─────
                    9 0            8 4
```

Add. Make 10 when possible.

14. 2 3   15. 1 2
     2 6   5 9
    +3 7   +3 1

16. 1 5   17. 3 4
     1 5   5 6
     4 4   1 1
    +2 4   + 9

18. There was a pie-eating contest at the fair. The numbers of pies eaten by the different contestants were 3, 5, 4, 5, 6, and 2. How many pies were eaten in all?

SYSTEMATIC REVIEW

11E

Skip count and write the numbers.

1. ___, 6, ___, ___, ___, 18, ___, ___, ___, ___

Skip count and write the missing numbers. Then fill in the missing factors under the lines.

2.

$\dfrac{0}{5 \cdot \_\_}$ $\dfrac{\_\_}{5 \cdot 1}$ $\dfrac{\_\_}{5 \cdot \_\_}$ $\dfrac{15}{5 \cdot \_\_}$ $\dfrac{\_\_}{5 \cdot \_\_}$ $\dfrac{\_\_}{5 \cdot 5}$

$\dfrac{\_\_}{5 \cdot \_\_}$ $\dfrac{\_\_}{5 \cdot 7}$ $\dfrac{\_\_}{5 \cdot \_\_}$ $\dfrac{\_\_}{5 \cdot 9}$ $\dfrac{50}{5 \cdot \_\_}$

Solve for the unknown.

3. $7D = 63$

4. $6F = 30$

5. $2F = 8$

6. $6B = 60$

Multiply.

7. $(3)(5) = $ ___

8. $8 \times 9 = $ ___

9. $9 \times 5 = $ ___

10. $8 \cdot 1 = $ ___

SYSTEMATIC REVIEW 11E

Find the area of each rectangle.

11. 8 in, 5 in A = ____

12. 2 mi, 2 mi A = ____

13. 9 in, 3 in A = ____

Add. Make 10 when possible.

14. 45
     31
    +15

15. 26
     84
    +32

16. 23
     15
     17
    + 2

17. 31
     29
     32
    + 8

18. Jamie had nine dimes. His dad gave him two nickels. How much money does Jamie have now?

19. Shawn drew 6 triangles. How many sides did he draw? (skip count)

20. Mrs. Post drove to town in 35 minutes on Monday. On Tuesday, there was a lot of traffic, and it took her 52 minutes to get to town. How many minutes longer did it take her to get to town on Tuesday?

SYSTEMATIC REVIEW 11F

Skip count and write the numbers.

1. ____, ____, 9, ____, ____, ____, ____, 24, ____, ____

Skip count and write the missing numbers. Then fill in the missing factors under the lines.

2.

$\dfrac{0}{2\times\underline{}}\quad \dfrac{}{2\times 1}\quad \dfrac{}{2\times\underline{}}\quad \dfrac{6}{2\times\underline{}}\quad \dfrac{}{2\times\underline{}}\quad \dfrac{}{2\times 5}$

$\dfrac{}{2\times\underline{}}\quad \dfrac{}{2\times 7}\quad \dfrac{}{2\times\underline{}}\quad \dfrac{}{2\times 9}\quad \dfrac{20}{2\times\underline{}}$

Solve for the unknown.

3. 2D = 6   4. 4F = 40

5. 6F = 54   6. 5B = 25

Multiply.

7. (9)(9) = ____   8. 4 × 9 = ____

9. 5 × 7 = ____   10. 6 • 0 = ____

SYSTEMATIC REVIEW 11F

Find the area of each rectangle.

11. 9 ft, 7 ft   A = _____

12. 1 ft, 1 ft   A = _____

13. 6 in, 2 in   A = _____

Add. Make 10 when possible.

14. 1 9
     9 1
   +   7

15. 1 7
     3 6
   + 4 4

16. 5 5
     4 1
     6 5
   +   2

17. 5 1
     1 2
     2 4
   + 3 8

18. A snack costs 45 cents. How many nickels are needed?

19. Seven dominos each show a three-dot pattern. How many dots are on all the dominos? (skip count)

20. Roger gave the cashier $50 to pay for shoes that cost $38. How many dollars should he get in change?

    He gave his change to his three children so that each one got the same amount. How much did each one receive?

APPLICATION AND ENRICHMENT

11G

Here is another Venn diagram to complete.

1. In region A, write the numbers that you say when you skip count by 3 from 3 to 30.

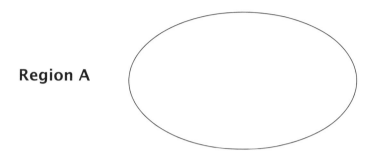

Region A

2. In region B, write the numbers that you say when you skip count by 9 from 9 to 90.

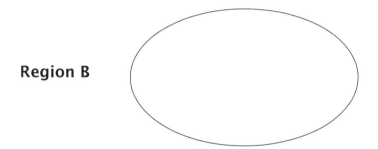

Region B

3. Now put the shapes together to make a Venn diagram. The numbers that are in both regions go in the middle where A and B overlap. The other numbers go in the outside part of the regions where they belong.

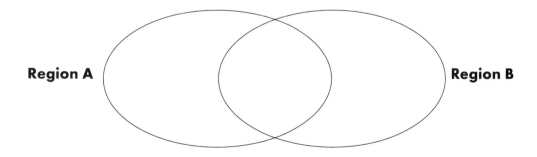

Region A   Region B

4. How many numbers are in the overlapping part of the shapes?

APPLICATION AND ENRICHMENT 11G

Color the picture. Complete each step in the order given for best results. If you have already colored a number, do not color it again in the next step.

If you say the number when you skip count by 9 to 90, color the space red.
If you say the number when you skip count by 3 to 30, color the space blue.
If you say the number when you skip count by 5 to 50, color the space green.
If you say the number when you skip count by 2 to 20, color the space yellow.

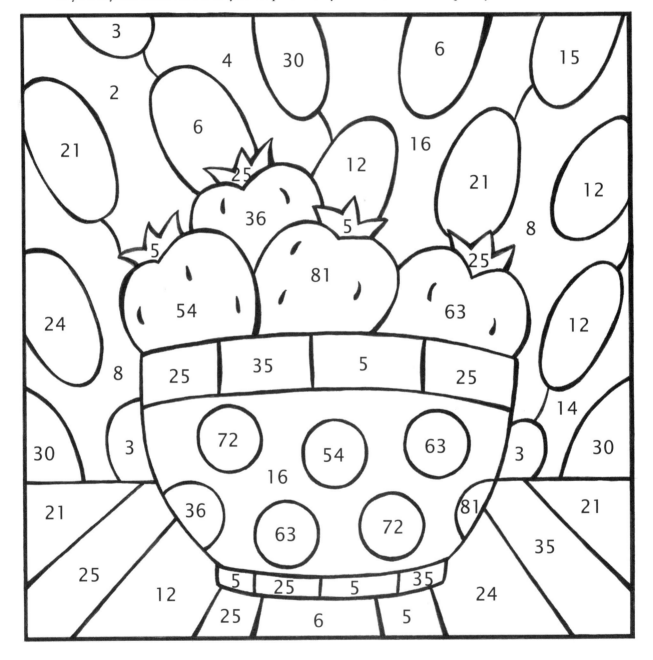

LESSON PRACTICE

12A

Find the answer by multiplying.

1. $3 \times 4 =$ _____

2. $9 \times 3 =$ _____

3. $3 \cdot 7 =$ _____

4. $10 \cdot 3 =$ _____

5. $3 \times 7 =$ _____

6. $1 \times 3 =$ _____

7. $(3)(4) =$ _____

8. $(8)(3) =$ _____

9. 8
   $\underline{\times\,3}$

10. 3
    $\underline{\times\,5}$

11. 3
    $\underline{\times\,3}$

12. 6
    $\underline{\times\,3}$

13. Three counted two times equals _____ .

14. Three counted six times equals _____ .

15. Use your blocks to build the numbers you say when skip counting by 3.

LESSON PRACTICE 12A

Skip count and write the missing numbers. Then fill in the missing factors under the lines.

16.
$$\frac{0}{(3)(0)} \quad \frac{\_\_}{(3)(\_\_)} \quad \frac{6}{(3)(\_\_)} \quad \frac{\_\_}{(3)(3)} \quad \frac{\_\_}{(3)(\_\_)} \quad \frac{\_\_}{(3)(5)}$$

$$\frac{\_\_}{(3)(\_\_)} \quad \frac{\_\_}{(3)(7)} \quad \frac{24}{(3)(\_\_)} \quad \frac{\_\_}{(3)(9)} \quad \frac{30}{(3)(\_\_)}$$

Find the missing multiples of 3 and 5 in the equivalent fractions.

17. $\dfrac{3}{5} = \dfrac{\_\_}{\_\_} = \dfrac{9}{\_\_} = \dfrac{\_\_}{20} = \dfrac{15}{30} = \dfrac{\_\_}{\_\_} = \dfrac{24}{\_\_} = \dfrac{\_\_}{\_\_} = \dfrac{30}{\_\_}$

Skip count or multiply by 3 to find how many feet. Use the symbol for feet when writing your answer.

18.

4 yards = _____

Skip count or multiply by 3 to find the number of teaspoons. Each spoon represents one tablespoon.

19.

6 tablespoons = _____ teaspoons

20. Mom's new bedspread is three yards long. How many feet long is the bedspread?_____

LESSON PRACTICE 12B

Find the answer by multiplying.

1. $3 \times 10 =$ _____
2. $3 \times 3 =$ _____

3. $3 \cdot 5 =$ _____
4. $6 \cdot 3 =$ _____

5. $3 \times 2 =$ _____
6. $0 \times 3 =$ _____

7. $(3)(9) =$ _____
8. $(7)(3) =$ _____

9. 1
   $\underline{\times\ 3}$

10. 3
    $\underline{\times\ 8}$

11. 4
    $\underline{\times\ 3}$

12. 10
    $\underline{\times\ 3}$

13. Three counted nine times equals _____ .

14. Three counted five times equals _____ .

15. $3 \times 6 =$ _____

    $6 \times 3 =$ _____

16. $3 \times 2 =$ _____

    $2 \times 3 =$ _____

LESSON PRACTICE 12B

Skip count or multiply by 3 to find the number of feet. Use the symbol for feet when writing your answer.

17.

| 1' | 1' | 1' | 1' | 1' | 1' | 1' | 1' | 1' | 1' | 1' | 1' |
| 1' | 1' | 1' | 1' | 1' | 1' | 1' | 1' | 1' | 1' | 1' | 1' |

8 yards = _____

Skip count or multiply by 3 to find the number of teaspoons.

18.

7 tablespoons = _____ teaspoons

19. Jill used four tablespoons of oil to cook the chicken. How many teaspoons did she use?

20. Martha bought five yards of ribbon to decorate pillows for the craft fair. She needs one foot of ribbon to trim each pillow. How many pillows can she trim?

LESSON PRACTICE

12C

Find the answer by multiplying.

1. $3 \times 1 =$ _____

2. $3 \times 5 =$ _____

3. $3 \cdot 10 =$ _____

4. $2 \cdot 3 =$ _____

5. $3 \times 3 =$ _____

6. $9 \times 3 =$ _____

7. $(3)(6) =$ _____

8. $(4)(3) =$ _____

9.   7
  $\times\ 3$

10.   3
  $\times\ 0$

11.   8
  $\times\ 3$

12.   5
  $\times\ 3$

13. Three counted eight times equals _____ .

14. Three counted three times equals _____ .

GAMMA LESSON PRACTICE 12C

LESSON PRACTICE 12C

Color the skip-count pattern for 3, going all the way to 99. Use the chart to find the answer to 3 × 12.

15.

0	1	2	3	4	5	6	7	8	9
10	11	12	13	14	15	16	17	18	19
20	21	22	23	24	25	26	27	28	29
30	31	32	33	34	35	36	37	38	39
40	41	42	43	44	45	46	47	48	49
50	51	52	53	54	55	56	57	58	59
60	61	62	63	64	65	66	67	68	69
70	71	72	73	74	75	76	77	78	79
80	81	82	83	84	85	86	87	88	89
90	91	92	93	94	95	96	97	98	99

Skip count or multiply by 3 to find the number of feet. Use the symbol for feet when writing your answer.

16.

| 1' | 1' | 1' | | 1' | 1' | 1' |

2 yd = _____

Skip count or multiply by 3 to find the number of teaspoons.

17.

9 Tbsp = _____ tsp

18. Leah used four tablespoons of sugar, one tablespoon of cinnamon, and one tablespoon of nutmeg in her pie. How many *teaspoons* of sugar and spice did she use in all?

SYSTEMATIC REVIEW 12D

Find the answer by multiplying.

1. 3 × 3 = _____
2. (8)(3) = _____

3. 3 • 7 = _____
4. 6 × 3 = _____

5. 　5
　× 9

6. 　6
　× 2

7. 　5
　× 5

8. 　8
　× 9

Fill in the blanks.

9. 4 qt = ___ pt
10. 7 nickels = ___ ¢

11. 5 Tbsp = ___ tsp
12. 6 yd = ___ ft

Add or subtract.

13. 　3 1
　　7 9
　　4 5
　+　 3

14. 　1 8
　　2 5
　　5 3
　+ 7 2

SYSTEMATIC REVIEW 12D

15. 69
    − 19
    ———

16. 81
    − 27
    ———

17. Sarah's room is 12 feet wide. Solve for the unknown and find the width of the room in yards.

18. Nine families in our church have three children each. How many children do they have in all?

19. Jessica's leaf collection consisted of 5 oak leaves, 13 maple leaves, 7 dogwood leaves, and 15 leaves from different kinds of fruit trees. How many leaves did Jessica have in her collection?

20. Scott had 8 dollars. He earned 18 more dollars mowing lawns. He spent 7 dollars on gasoline for the mower. How much money did he have left?

SYSTEMATIC REVIEW 12E

Find the answer by multiplying.

1. $3 \times 5 =$ _____

2. $(10)(3) =$ _____

3. $3 \cdot 6 =$ _____

4. $8 \times 3 =$ _____

5. 5
   $\underline{\times 8}$

6. 7
   $\underline{\times 9}$

7. 10
   $\underline{\times 6}$

8. 8
   $\underline{\times 2}$

Fill in the blanks.

9. 7 qt = ___ pt

10. 4 dimes = ___ ¢

11. 9 Tbsp = ___ tsp

12. 3 yd = ___ ft

Add or subtract.

13. 57
    21
    22
    $\underline{+15}$

14. 43
    44
    63
    $\underline{+11}$

15. 53
    −24

16. 65
    −19

17. Becky's bread recipe calls for 2 teaspoons of salt. She wants to make three times the recipe. How many teaspoons of salt does she need?

    How many tablespoons of salt does Becky need? (Solve for unknown.)

18. On the table in Mindy's kitchen are 3 baskets. Each basket has 3 apples. How many apples does she have?

    Each apple has 3 worms. How many worms are there?

19. One day as part of her training for the Olympics, Lisa ran 12 miles, rode her bike for 25 miles, swam for 3 miles, and sailed for 10 miles. How many miles did she cover?

20. The custodian polished 11 square yards of the hallway before lunch. Since the hallway is 3 yards wide and 7 yards long, how many square yards are left to polish? (Hint: First find the area of the hallway.)

SYSTEMATIC REVIEW

12F

Find the answer by multiplying.

1. $4 \times 3 =$ _____

2. $(7)(3) =$ _____

3. $3 \cdot 8 =$ _____

4. $1 \times 3 =$ _____

5. $1\,0$
   $\underline{\times3}$

6. 6
   $\underline{\times\,3}$

7. 9
   $\underline{\times\,6}$

8. 5
   $\underline{\times\,4}$

9. 9 qt = ___ pt

10. 6 nickels = ___ ¢

11. 3 Tbsp = ___ tsp

12. 5 yd = ___ ft

Add or subtract.

13. $1\,7$
    $1\,6$
    $1\,2$
    $\underline{+\,1\,4}$

14. $3\,5$
    $7\,4$
    $1\,5$
    $\underline{+\,2\,4}$

SYSTEMATIC REVIEW 12F

15. 41
    − 32

16. 38
    − 29

17. James was up to bat. He hit the ball, and it flew a total of 30 feet before it was caught. Solve for the unknown to find how many yards the ball went.

18. Julie had $30 dollars to spend on knitting yarn. She bought three very small balls of cashmere yarn at $9 apiece. How much money did she have left?

19. Mark's dog adoption center was home to 57 big dogs and 14 little dogs. If 71 of the dogs were adopted, how many dogs were left?

20. After counting the money in her pocket, Christine found she had 4 dimes, 3 nickels, and 11 pennies. How much money did Christine have in her pocket?

12G

APPLICATION AND ENRICHMENT

Multiply each number in the top row by three. Write the answers in the bottom row.

0	1	2	3	4	5	6	7	8	9	10
0	3									

Look at the numbers in the bottom row. Start at 0 on the circle and connect the dots in order. Use the numbers in the units place of each answer in the bottom row.

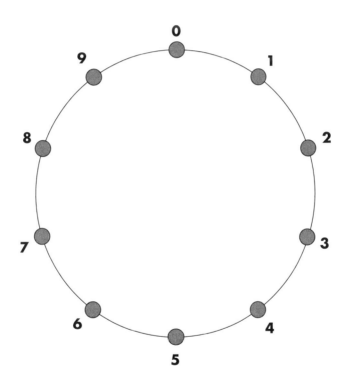

1. What shape did you make?

2. How many points does your shape have?

APPLICATION AND ENRICHMENT 12G

Skip count by three. Start at the star and connect the dots all the way to 99. You may use the chart on 12C in the student book if you need help. Use the picture to practice skip counting by three.

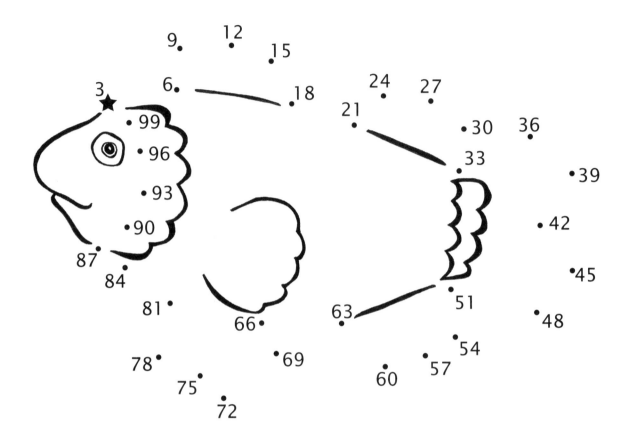

Gloria Goldfish ignored three worms on hooks every day.

How many worms on hooks did she ignore in seven days?

_____

LESSON PRACTICE 13A

Skip count by 6. Write the correct numbers in the squares with lines.

1. 6, __, __, __, 30, __, __, __, 54, __

2. __, 12, __, __, __, __, __, 48, __, __

LESSON PRACTICE 13A

Skip count and write the numbers.

3. 6, ____, ____, ____, ____, ____, ____, ____, ____, 60

Fill in the blanks in the numerators and denominators to name the equivalent fractions.

4.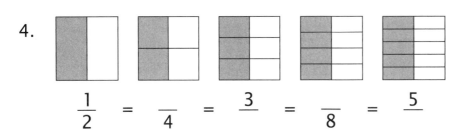

$$\frac{1}{2} = \frac{}{4} = \frac{3}{} = \frac{}{8} = \frac{5}{}$$

5.

$$\frac{1}{3} = \frac{}{6} = \frac{3}{} = \frac{4}{12} = \frac{}{}$$

Use skip counting to solve the word problems.

6. Rowena painted three paintings a week for six weeks. How many paintings did she paint altogether?

7. The preschool class had eight children. Each child ate six crackers. How many crackers were eaten?

LESSON PRACTICE 13B

Skip count by 6. Write the correct numbers in the squares with lines.

1.

					18
					60

2.

					42

LESSON PRACTICE 13B

Skip count and write the numbers.

3. ____, 12, ____, ____, ____, ____, ____, 48, ____, ____

Fill in the blanks in the numerators and denominators to name the equivalent fractions.

4.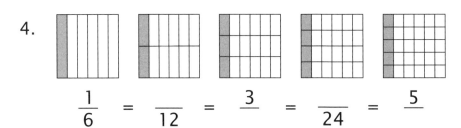

$$\frac{1}{6} = \frac{}{12} = \frac{3}{} = \frac{}{24} = \frac{5}{}$$

5.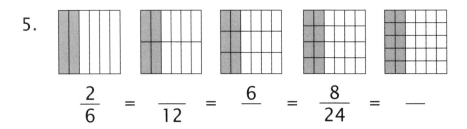

$$\frac{2}{6} = \frac{}{12} = \frac{6}{} = \frac{8}{24} = \frac{}{}$$

Use skip counting to solve the word problems.

6. Thomas ate two ice cream cones a day for six days. How many ice cream cones did he eat?

7. We were setting tables for the banquet. Each table seated six people. We had to set six tables to seat everyone. How many people were we expecting?

LESSON PRACTICE 13C

Skip count by 6. Write the correct numbers in the squares with lines.

1.

2.

Skip count and write the numbers.

3. ____, ____, ____, ____, ____, ____, ____, ____, ____, 60

Fill in the blanks in the numerators and denominators to name the equivalent fractions.

4.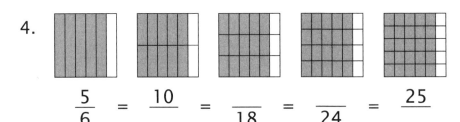

$$\frac{5}{6} = \frac{10}{} = \frac{}{18} = \frac{}{24} = \frac{25}{}$$

5.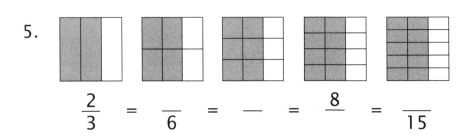

$$\frac{2}{3} = \frac{}{6} = \frac{}{} = \frac{8}{} = \frac{}{15}$$

Use skip counting to solve the word problems.

6. A pilot took his plane up six times a day. How many times did he take it up in nine days?

7. The teacher saw six hands raised in the air. How many fingers were raised?

SYSTEMATIC REVIEW 13D

Skip count and write the numbers.

1. ____, 12, ____, ____, ____, ____, ____, 48, ____, ____

Fill in the blanks in the numerators and denominators to name the equivalent fractions.

2.

$$\frac{3}{5} = \frac{}{} = \frac{}{15} = \frac{12}{} = \frac{}{}$$

Multiply.

3. (9)(3) = _____

4. 2 × 6 = _____

5. 3 × 4 = _____

6. 5 • 5 = _____

7. 8
   × 9

8. 6
   × 5

9. 3
   × 7

10. 1 0
    × 8

Add or subtract.

11. 7 3
    + 4 5

12. 3 8
    + 6 7

SYSTEMATIC REVIEW 13D

13. 54
    − 25

14. 88
    − 19

QUICK REVIEW

Perimeter is the distance around a figure. It is found by adding the lengths of each side. Remember that area and perimeter are different. For example, the perimeter of a backyard tells you the length of fence that is needed to enclose it. The area of the yard helps you find how much grass seed is needed to cover the entire space inside the fence.

Find the perimeter of each rectangle or square. The first one has been done for you.

15. Rectangle: 9" top, 9" bottom, 5" left, 5" right.
P = 5 + 9 + 5 + 9 = 28 in

16. Square: 3 mi on each side.
P = _____

17. Rectangle: 12' top, 12' bottom, 10' left, 10' right.
P = _____

18. Jeff has a square garden which measures 25 feet on a side. How many feet of fence does he need to enclose his garden?

19. Each bird laid six eggs in its nest. There were six birds. How many eggs were in all the nests?

20. Molly bought six yards of fabric for curtains. How many feet of fabric did she buy?

SYSTEMATIC REVIEW 13E

Skip count and write the numbers.

1. ___, ___, 18, ___, ___, ___, ___, ___, 54, ___

Fill in the blanks in the numerators and denominators to name the equivalent fractions.

2. $\frac{2}{5} = \frac{}{} = \frac{}{15} = \frac{8}{} = \frac{}{}$

Multiply.

3. $(2)(7) =$ _____

4. $3 \times 0 =$ _____

5. $9 \times 9 =$ _____

6. $5 \cdot 3 =$ _____

7. 3
   $\underline{\times 3}$

8. 9
   $\underline{\times 4}$

9. 10
   $\underline{\times 7}$

10. 8
    $\underline{\times 2}$

SYSTEMATIC REVIEW 13E

Add or subtract.

11. 6 1
    + 2 2

12.   4 5
     +   9

13. 7 6
    − 3 8

14. 9 3
    − 4 4

Find the perimeter of each rectangle.

15. Rectangle: 16 yd (top), 20 yd (left), 20 yd (right), 16 yd (bottom). P = _____

16. Rectangle: 7' (top), 4' (left), 4' (right), 7' (bottom). P = _____

17. Square: 5" on each side. P = _____

18. A city is 6 miles long and 4 miles wide. How long is the road that goes all around the boundaries of the city?

19. Fencing comes in six-foot lengths. How many feet of fence will Jeff have if he buys eight lengths?

20. Mom used eight tablespoons of honey in the bread she was making. How many teaspoons of honey did she use?

SYSTEMATIC REVIEW 13F

Skip count and write the numbers.

1. 9, ___, ___, ___, ___, ___, ___, ___, ___, ___

Fill in the blanks in the numerators and denominators to name the equivalent fractions.

2.

$$\frac{3}{6} = \frac{}{} = \frac{}{} = \frac{}{24} = \frac{15}{}$$

Multiply.

3. (3)(9) = ___

4. 9 × 2 = ___

5. 2 × 4 = ___

6. 10 • 6 = ___

7. 9 × 5

8. 5 × 4

9. 9 × 6

10. 7 × 9

SYSTEMATIC REVIEW 13F

Add or subtract.

11. 17
    +18

12. 34
    +79

13. 48
    −27

14. 53
    −19

Find the perimeter of each rectangle.

15. 11 mi (top), 14 mi (left), 14 mi (right), 11 mi (bottom)
P = _____

16. 21' (top), 12' (left), 12' (right), 21' (bottom)
P = _____

17. 10" (top), 10" (left), 10" (right), 10" (bottom)
P = _____

18. Ashley drew a rectangle that was five inches long and one inch wide. What is the perimeter of the rectangle?

19. What is the area of the rectangle in #18?

20. There are five fingers on a hand and two hands on a person. How many fingers are there on nine people?

APPLICATION AND ENRICHMENT

13G

Help to write the word problems. Fill in the blanks and multiply to find the answers. Write your answers in the boxes.

1. Jacob caught nine _____ every day.

   How many did he catch in six days?

2. Isabella made five _____ an hour.

   How many could she make in six hours?

3. Noah ate three _____ at each meal.

   How many did he eat at six meals?

4. Emma saw two _____ in every tree.

   There were six trees. How many did Emma see in all the trees?

APPLICATION AND ENRICHMENT 13G

Use the chart to keep track of the weather for a week. Color one box each day to tell if it is sunny, cloudy, raining, or snowing.

Sunny	**Cloudy**	**Rain or Snow**

How many days were sunny?

How many days were cloudy?

How many days had rain or snow?

LESSON PRACTICE 14A

Find the answer by multiplying.

1. $6 \times 9 =$ _____
2. $7 \times 6 =$ _____
3. $10 \times 6 =$ _____
4. $4 \times 6 =$ _____
5. $(5)(6) =$ _____
6. $(1)(6) =$ _____
7. $(6)(3) =$ _____
8. $(6)(2) =$ _____
9. $9 \cdot 6 =$ _____
10. $6 \cdot 4 =$ _____
11. $6 \cdot 8 =$ _____
12. $7 \cdot 6 =$ _____

Find the answer by multiplying.

13. 6
    × 4
    ---

14. 6
    × 1
    ---

15. 6
    × 5
    ---

16. 6
    × 3

GAMMA LESSON PRACTICE 14A

LESSON PRACTICE 14A

17. Six counted eight times equals _____ .

18. Six counted six times equals _____ .

Color all the boxes that have a number you would say when skip counting by 6. Finish the pattern, and see whether you can tell the answers for 6 × 11 and 6 × 12.

19.
0	1	2	3	4	5	6	7	8	9
10	11	12	13	14	15	16	17	18	19
20	21	22	23	24	25	26	27	28	29
30	31	32	33	34	35	36	37	38	39
40	41	42	43	44	45	46	47	48	49
50	51	52	53	54	55	56	57	58	59
60	61	62	63	64	65	66	67	68	69
70	71	72	73	74	75	76	77	78	79
80	81	82	83	84	85	86	87	88	89
90	91	92	93	94	95	96	97	98	99

20. Ants have six legs each. Seven ants are working in a tunnel. How many legs are working?

LESSON PRACTICE 14B

Find the answer by multiplying.

1. 6 × 0 = _____
2. 6 × 6 = _____
3. 2 × 6 = _____
4. 8 × 6 = _____
5. (3)(6) = _____
6. (5)(6) = _____
7. (6)(9) = _____
8. (6)(10) = _____
9. 6 · 6 = _____
10. 6 · 1 = _____
11. 6 · 4 = _____
12. 2 · 6 = _____

Find the answer by multiplying.

13. 6
   × 7

14. 6
   × 5

15. 6
   × 3

16. 9
   × 6

LESSON PRACTICE 14B

17. Use your blocks to build the numbers you say when skip counting by 6.

Skip count and write the missing numbers. Then fill in the missing factors under the lines.

18.

$\dfrac{0}{6\cdot 0} \quad \dfrac{}{6\cdot \underline{}} \quad \dfrac{}{6\cdot \underline{}} \quad \dfrac{}{6\cdot 3} \quad \dfrac{24}{6\cdot \underline{}} \quad \dfrac{}{6\cdot \underline{}}$

$\dfrac{}{6\cdot \underline{}} \quad \dfrac{}{6\cdot 7} \quad \dfrac{}{6\cdot \underline{}} \quad \dfrac{}{6\cdot \underline{}} \quad \dfrac{60}{6\cdot \underline{}}$

19. Six counted 10 times equals _____ .

20. There are six red cars in the parking lot. What is the total number of wheels on those cars?

LESSON PRACTICE

Find the answer by multiplying.

1. 6 × 10 = _____

2. 6 × 8 = _____

3. 6 × 6 = _____

4. 4 × 6 = _____

5. (7)(6) = _____

6. (9)(6) = _____

7. (6)(5) = _____

8. (6)(3) = _____

9. 6 · 0 = _____

10. 6 · 6 = _____

11. 6 · 2 = _____

12. 8 · 6 = _____

Find the answer by multiplying.

13. 6
   × 5
   ___

14. 4
   × 6
   ___

15. 6
   × 1
   ___

16. 7
   × 6
   ___

LESSON PRACTICE 14C

Skip count and write the missing numbers. Then fill in the missing factors under the lines.

17.

$$\frac{0}{6\times 0} \quad \frac{\_\_\_}{\_\_\_} \quad \frac{\_\_\_}{6\times 2} \quad \frac{\_\_\_}{\_\_\_} \quad \frac{\_\_\_}{\_\_\_} \quad \frac{\_\_\_}{6\times 5}$$

$$\frac{\_\_\_}{\_\_\_} \quad \frac{\_\_\_}{\_\_\_} \quad \frac{48}{\_\_\_} \quad \frac{\_\_\_}{\_\_\_} \quad \frac{\_\_\_}{6\times 10}$$

18. Six counted nine times equals _____ .

19. Six insects landed on my window. I could see all of their feet. If insects have six legs apiece, how many insect feet did I see on my window?

20. The six houses on our road each have six apartments. How many apartments are on our road?

SYSTEMATIC REVIEW

14D

Multiply.

1. (10)(7) = _____

2. 9 × 6 = _____

3. 5 × 0 = _____

4. 6 · 3 = _____

5.   3
  × 9
  ____

6.   5
  × 4
  ____

7.   4
  × 6
  ____

8.   7
  × 9
  ____

9.   6
  × 5
  ____

10.   6
  × 6
  ____

11.   9
  × 5
  ____

12.   7
  × 2
  ____

Find the perimeter of each shape.

13.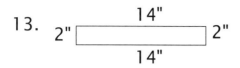

P = _____

14.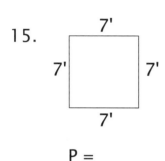

P = _____

15.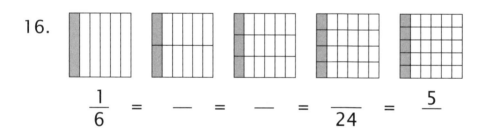

P = _____

Fill in the blanks in the numerators and denominators to name the equivalent fractions.

16. $\dfrac{1}{6} = \dfrac{\_\_}{\_\_} = \dfrac{\_\_}{\_\_} = \dfrac{\_\_}{24} = \dfrac{5}{\_\_}$

17. Lois drank six glasses of water every day last week (seven days). How many glasses of water did she drink in all?

18. Devan has five dimes, three nickels, and eight pennies. How much money does he have?

19. There were 50 apples hanging on a tree. If 23 of them were rotten, how many good apples were on the tree?

20. On the way to the mountain, four people rode in the car, and eight rode in the van. On the way home, six people rode in the car. How many were left to ride in the van?

SYSTEMATIC REVIEW

Multiply.

1. $(3)(9) = $ _____

2. $6 \times 9 = $ _____

3. $5 \times 5 = $ _____

4. $9 \cdot 8 = $ _____

5. $\begin{array}{r} 4 \\ \times\ 3 \\ \hline \end{array}$

6. $\begin{array}{r} 6 \\ \times\ 8 \\ \hline \end{array}$

7. $\begin{array}{r} 9 \\ \times\ 0 \\ \hline \end{array}$

8. $\begin{array}{r} 7 \\ \times\ 5 \\ \hline \end{array}$

9. $\begin{array}{r} 3 \\ \times\ 6 \\ \hline \end{array}$

10. $\begin{array}{r} 9 \\ \times\ 9 \\ \hline \end{array}$

11. $\begin{array}{r} 1 \\ \times\ 8 \\ \hline \end{array}$

12. $\begin{array}{r} 6 \\ \times\ 7 \\ \hline \end{array}$

Find the perimeter of each shape.

13.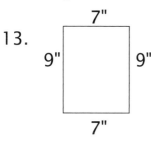

   P = _____

14.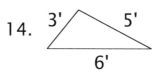

   P = _____

15. 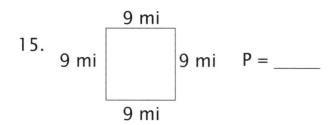 P = _____

Fill in the blanks in the numerators and denominators to name the equivalent fractions.

16.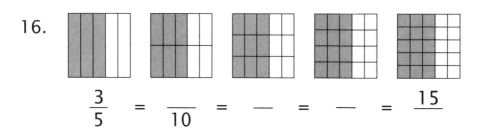

$$\frac{3}{5} = \frac{}{10} = \frac{}{} = \frac{}{} = \frac{15}{}$$

17. Wayne made a peg board. He hammered eight rows of nails with six nails in each row. How many nails did he use?

18. Erin made cards. She made 4 gift cards, 16 birthday cards, 7 Valentine cards, 19 congratulations cards, and 6 get-well cards. How many cards did she make?

19. A room is 10 yards long. How many feet long is the room?

20. The room in #19 is a rectangle that is 21 feet wide. What is the perimeter of the room?

SYSTEMATIC REVIEW

14F

Multiply.

1. (4)(9) = _____

2. 3 × 3 = _____

3. 6 × 6 = _____

4. 7 · 6 = _____

5.   9
  × 7

6.   3
  × 8

7.   9
  × 5

8.   4
  × 6

9.   8
  × 2

10.   9
  × 3

11.   6
  × 3

12.   6
  × 5

Find the perimeter of each shape.

13.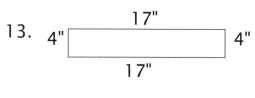

P = _____

14.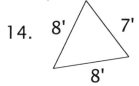

P = _____

15. 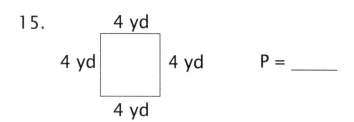   P = _____

Fill in the blanks in the numerators and denominators to name the equivalent fractions.

16.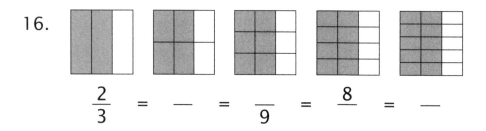

$$\frac{2}{3} = \frac{}{} = \frac{}{9} = \frac{8}{} = \frac{}{}$$

17. A recipe calls for five tablespoons of honey. How many teaspoons are needed?

18. How many nickels would Jenna need to pay for a candy bar that costs 20 cents?

19. Three dogs and seven children walked through the fresh snow. How many feet were making tracks?

20. Cameron cut a piece of wrapping paper that measures 3 feet wide and 6 feet long. What is the area of the paper?

APPLICATION AND ENRICHMENT

14G

Multiply each number in the top row by six. Write the answers in the bottom row.

0	1	2	3	4	5	6	7	8	9	10
0	6									

Look at the numbers in the bottom row. Start at 0 on the square and connect the dots in order, using the numbers in the units place of each answer in the bottom row.

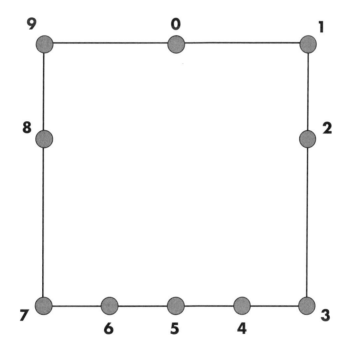

1. What shape did you make?

2. How many points does your shape have?

APPLICATION AND ENRICHMENT 14G

Skip count by six. Start at the star and connect the dots all the way to 96. You may use the chart on 14A in the student book to help you if needed. Use the picture to practice skip counting by six.

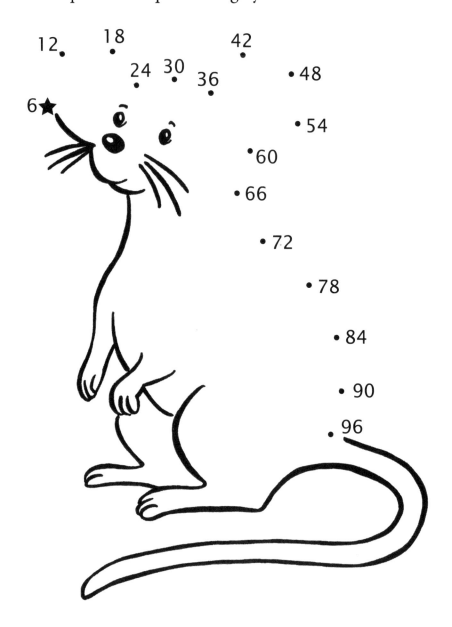

Mary Mouse likes to have six nibbles of cheese for each meal.

How many nibbles of cheese would she need for 12 meals?

Use the numbers on the picture to help you find the answer.

_____

LESSON PRACTICE 15A

Skip count by 4. Write the correct numbers in the squares with lines.

1.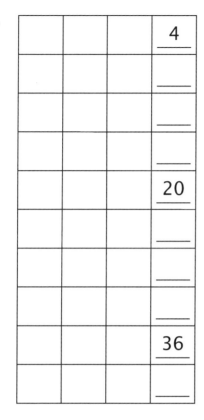

2.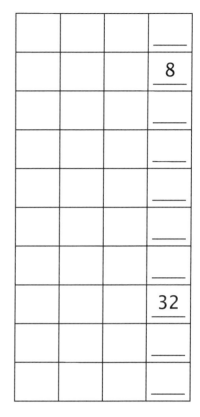

Skip count and write the numbers.

3. 4, ____, ____, ____, ____, ____, ____, ____, ____, ____

Find the missing multiples of 4 and 5 in the equivalent fractions.

4. $\dfrac{4}{5} = \dfrac{}{} = \dfrac{12}{} = \dfrac{}{20} = \dfrac{}{} = \dfrac{}{} = \dfrac{}{35} = \dfrac{32}{} = \dfrac{}{} = \dfrac{40}{}$

LESSON PRACTICE 15A

Skip count by 4 to find the number of quarts in all.

5.

Use skip counting to solve the word problems.

6. Mom bought four pairs of socks for each of her three children. How many socks did she buy?

7. Each patio table by the pool was surrounded by four chairs. How many people could be seated at nine tables?

8. Tom bought five gallons of milk in quart jugs. How many jugs of milk did he bring home?

LESSON PRACTICE 15B

Skip count by 4. Write the correct numbers in the squares with lines.

1._____   2._____

(grids for skip counting by 4; given values: 4, 16 in #1; 12, 36 in #2)

Skip count and write the numbers.

3. ____, 8, ____, ____, ____, ____, ____, ____, ____, 40

Find the missing multiples of 4 and 9 in the equivalent fractions.

4. $\dfrac{4}{9} = \dfrac{}{} = \dfrac{}{} = \dfrac{16}{} = \dfrac{}{} = \dfrac{}{54} = \dfrac{28}{} = \dfrac{}{} = \dfrac{}{} = \dfrac{}{}$

Skip count by 4 to find the number of quarts in all.

5.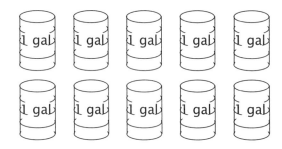

Use skip counting to solve the word problems.

6. How many tires are needed to get four cars ready to drive?

7. Four laps around the track is equal to one mile. Sherrie ran six miles. How many laps did she run?

8. Two gallons of oil are needed for the monster truck. How many quarts of oil should the mechanic buy?

LESSON PRACTICE 15C

Skip count by 4. Write the correct numbers in the squares with lines.

1.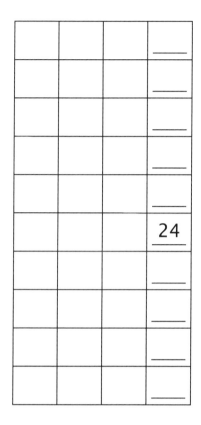

2.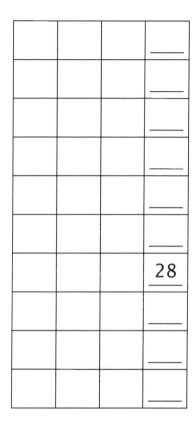

Skip count and write the numbers.

3. ____, ____, ____, 16, ____, ____, ____, ____, 36, ____

Find the missing multiples of 2 and 4 in the equivalent fractions.

4. $\dfrac{2}{4} = \dfrac{}{} = \dfrac{}{} = \dfrac{}{20} = \dfrac{12}{} = \dfrac{}{} = \dfrac{}{} = \dfrac{}{} = \dfrac{}{}$

Skip count by 4 to find the number of quarts in all.

5.

Use skip counting to solve the word problems.

6. Daniel took four pictures every minute. How many pictures did he take in three minutes?

7. Mary planted eight bulbs in her garden. Two years later, each plant had four flowers. How many flowers did Mary have altogether in her garden?

8. After making nine gallons of tomato juice, Sadie found she had only quart jars to put it in. How many quart jars will she need for the tomato juice?

SYSTEMATIC REVIEW 15D

Skip count and write the numbers.

1. ____, 8, ____, 16, ____, ____, ____, ____, ____, ____

Fill in the blanks in the numerators and denominators to name the equivalent fractions.

2.

$$\frac{3}{4} = \frac{}{} = \frac{}{} = \frac{}{16} = \frac{15}{}$$

Multiply.

3. (10)(7) = _____

4. 9 × 6 = _____

5. 5 × 0 = _____

6. 6 · 3 = _____

7.   9
  × 3

8.   5
  × 4

9.   6
  × 4

10.   7
  × 9

Solve for the unknown.

11. 5K = 30

12. 6A = 18

13. 9X = 45

14. 6B = 36

Find the perimeter and area of each rectangle. The first one had been done for you.

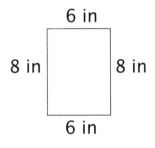

6 × 8 = 48
6 + 8 + 6 + 8 = 28

15. Area = __48 sq in__

17. Area = _____

16. Perimeter = __28 in__

18. Perimeter = _____

19. Amy placed two rings on each of four fingers on her left hand. How many rings is Amy wearing?

20. Four gallons of punch are needed for the party. The ingredients for the punch are all sold in quart containers. How many quarts should be purchased? (skip count)

SYSTEMATIC REVIEW

15E

Skip count and write the numbers.

1. ____, 6, ____, 12, ____, ____, ____, ____, ____, ____

Fill in the blanks in the numerators and denominators to name the equivalent fractions.

2. $\dfrac{1}{4} = \dfrac{}{} = \dfrac{}{12} = \dfrac{4}{} = \dfrac{}{}$

Multiply.

3. (2)(7) = _____

4. $9 \times 3 =$ _____

5. $9 \times 6 =$ _____

6. $5 \cdot 5 =$ _____

7. 9
   $\times\,8$

8. 4
   $\times\,3$

9. 6
   $\times\,8$

10. 9
    $\times\,0$

SYSTEMATIC REVIEW 15E

Solve for the unknown.

11. $9Y = 0$

12. $7G = 35$

13. $8R = 72$

14. $9T = 27$

Find the perimeter and area of each rectangle.

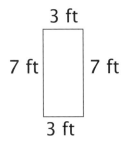

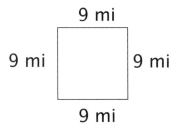

15. A = _____

16. P = _____

17. A = _____

18. P = _____

19. Four flies walked up the wall. How many feet were strolling on the wall? (skip count)

20. Peter threw a ball four yards, Arnold threw it six yards, and Gerry threw it nine yards. How many feet did they throw the ball in all? (Change each distance to feet and then add.)

SYSTEMATIC REVIEW 15F

Skip count and write the numbers.

1. ____, 8, ____, ____, ____, ____, ____, 32, ____, ____

Fill in the blanks in the numerators and denominators to name the equivalent fractions.

2. $\frac{2}{5} = \frac{}{} = \frac{}{} = \frac{8}{20} = \frac{}{}$

Multiply.

3. $(8)(1) = $ _____

4. $7 \times 6 = $ _____

5. $4 \times 9 = $ _____

6. $3 \cdot 3 = $ _____

7. 6
   $\times 6$

8. 7
   $\times 6$

9. 7
   $\times 9$

10. 3
    $\times 8$

Solve for the unknown.

11. 3X = 18

12. 6B = 24

13. 2E = 16

14. 9Y = 81

Find the perimeter and area of each rectangle.

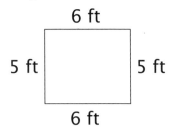

15. A = _____

16. P = _____

17. A = _____

18. P = _____

19. One week has seven days. How many days are in four weeks? (skip count)

20. Kelly bought 3 quarts of orange juice, 7 quarts of apple juice, 2 quarts of cranberry juice, and 5 quarts of grape juice. How many *pints* of juice did Kelly bring home?

APPLICATION AND ENRICHMENT

15G

On the chart, write the numbers that you say when you skip count by four. Continue to count all the way to 48.

0	1	2	3		5	6	7		9
10	11		13	14	15		17	18	19
	21	22	23		25	26	27		29
30	31		33	34	35		37	38	39
	41	42	43		45	46	47		49

1. How many numbers did you write?

Find all the numbers that you could say when you skip count by two. Color the boxes yellow.

2. How many numbers on the chart are answers to both two facts and four facts?

Find all the numbers that you could say when you skip count by three. Color the boxes blue. The numbers that are answers to both three facts and two facts should look green on the chart.

3. How many numbers on the chart are answers to both three facts and two facts?

APPLICATION AND ENRICHMENT 15G

Here are some more logic problems with Carroll diagrams to help you solve them. If you need to review these, look back at the logic problems on the application and enrichment pages for lesson 8.

1. Anna, Emily, and James all went on different vacations. Anna does not like heights. Emily is allergic to animals. James was bitten by a crab. Where did each person go for vacation?

	mountains	horse ranch	beach
Anna			
Emily			
James			

2. Alex, Jayna, and Billy all got to the picnic in different ways. Nobody drove Alex. Jayna's mom does not have a driver's license. How did each person get to the picnic?

	bus	Mom's car	walked
Alex			
Jayna			
Billy			

LESSON PRACTICE 16A

Find the answer by multiplying.

1. $4 \times 9 =$ _____

2. $7 \times 4 =$ _____

3. $10 \cdot 4 =$ _____

4. $4 \cdot 6 =$ _____

5. $5 \times 4 =$ _____

6. $1 \times 4 =$ _____

7. $(4)(3) =$ _____

8. $(4)(2) =$ _____

9. 4
   $\times\, 4$

10. 4
    $\times\, 8$

11. 4
    $\times\, 5$

12. 6
    $\times\, 4$

13. Four counted nine times equals _____ .

14. Four counted four times equals _____ .

GAMMA   LESSON PRACTICE 16A

LESSON PRACTICE 16A

Color all the boxes that have a number you would say when skip counting by 4. Finish the pattern and see whether you can tell the answer for 4×11 and 4×12.

15.
0	1	2	3	4	5	6	7	8	9
10	11	12	13	14	15	16	17	18	19
20	21	22	23	24	25	26	27	28	29
30	31	32	33	34	35	36	37	38	39
40	41	42	43	44	45	46	47	48	49
50	51	52	53	54	55	56	57	58	59
60	61	62	63	64	65	66	67	68	69
70	71	72	73	74	75	76	77	78	79
80	81	82	83	84	85	86	87	88	89
90	91	92	93	94	95	96	97	98	99

Multiply by 4 to find that number of quarters that are the same as six dollars.

16.

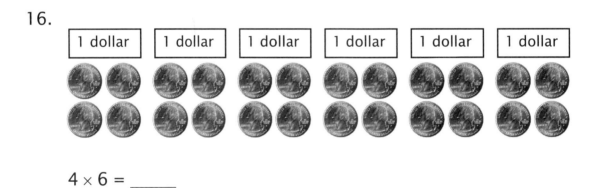

$4 \times 6 =$ _____

17. How many corners are in seven squares?

18. Ada needs quarters to wash her clothes. She put three dollar bills in the change machine. How many quarters did she get?

LESSON PRACTICE

16B

Find the answer by multiplying.

1. $4 \times 0 =$ _____

2. $4 \times 10 =$ _____

3. $4 \cdot 3 =$ _____

4. $6 \cdot 4 =$ _____

5. $2 \times 4 =$ _____

6. $4 \times 4 =$ _____

7. $(4)(7) =$ _____

8. $(9)(4) =$ _____

9.   1
  × 4
  ____

10.   4
  × 5
  ____

11.   4
  × 8
  ____

12.   3
  × 4
  ____

13. Use your blocks to build the numbers you say when skip counting by 4.

LESSON PRACTICE 16B

Skip count and write the missing numbers. Then fill in the missing factors under the lines.

14.

$\dfrac{0}{(4)(0)}$ $\dfrac{}{(4)(\underline{\ \ })}$ $\dfrac{}{(4)(\underline{\ \ })}$ $\dfrac{}{(4)(3)}$ $\dfrac{16}{(4)(\underline{\ \ })}$ $\dfrac{}{(4)(\underline{\ \ })}$

$\dfrac{}{(4)(\underline{\ \ })}$ $\dfrac{}{(4)(7)}$ $\dfrac{}{(4)(\underline{\ \ })}$ $\dfrac{}{(4)(\underline{\ \ })}$ $\dfrac{40}{(4)(\underline{\ \ })}$

Multiply by 4 to find the number of quarters that are the same as four dollars.

15.

1 dollar   1 dollar   1 dollar   1 dollar

4 × 4 = _____

16. Four counted seven times equals _____ .

17. I need four plates for each table. How many plates do I need for nine tables?

18. How many horseshoes will the blacksmith need for five horses?

LESSON PRACTICE 16C

Find the answer by multiplying.

1. $4 \times 2 =$ _____

2. $4 \times 4 =$ _____

3. $6 \cdot 4 =$ _____

4. $4 \cdot 10 =$ _____

5. $1 \times 4 =$ _____

6. $4 \times 5 =$ _____

7. $(4)(7) =$ _____

8. $(3)(4) =$ _____

9. 8
   × 4

10. 4
    × 9

11. 4
    × 6

12. 4
    × 4

13. Use your blocks to build the numbers you say when skip counting by 4.

LESSON PRACTICE 16C

Skip count and write the missing numbers. Then fill in the missing factors under the lines.

14.

$\dfrac{0}{4\cdot 0}\ \dfrac{}{4\cdot\underline{}}\ \dfrac{}{4\cdot\underline{}}\ \dfrac{}{4\cdot\underline{}}\ \dfrac{}{4\cdot 4}\ \dfrac{20}{4\cdot\underline{}}$

$\dfrac{}{4\cdot\underline{}}\ \dfrac{}{4\cdot 7}\ \dfrac{}{4\cdot\underline{}}\ \dfrac{}{4\cdot\underline{}}\ \dfrac{40}{4\cdot 10}$

Multiply by 4 to find the number of quarters.

15.

| 1 dollar | 1 dollar | 1 dollar | 1 dollar |

| 1 dollar | 1 dollar | 1 dollar | 1 dollar |

$4 \times 8 =$ _____

16. Four counted six times equals _____

17. Judah bought paper with 12 quarters. Solve for the unknown to find the number of dollars he used to pay for the paper.

18. Aiden saved $10 in quarters. How many quarters did he have?

16D

SYSTEMATIC REVIEW

Multiply.

1. $(4)(4) = $ _____

2. $2 \times 6 = $ _____

3. $6 \times 3 = $ _____

4. $10 \cdot 3 = $ _____

5. $\begin{array}{r} 5 \\ \times\, 9 \\ \hline \end{array}$

6. $\begin{array}{r} 4 \\ \times\, 6 \\ \hline \end{array}$

7. $\begin{array}{r} 2 \\ \times\, 7 \\ \hline \end{array}$

8. $\begin{array}{r} 3 \\ \times\, 3 \\ \hline \end{array}$

9. $\begin{array}{r} 4 \\ \times\, 9 \\ \hline \end{array}$

10. $\begin{array}{r} 5 \\ \times\, 3 \\ \hline \end{array}$

11. $\begin{array}{r} 7 \\ \times\, 6 \\ \hline \end{array}$

12. $\begin{array}{r} 9 \\ \times\, 3 \\ \hline \end{array}$

SYSTEMATIC REVIEW 16D

Solve for the unknown. The first one has been done for you.

13. 32 quarters = __8__ dollars   4 × 8 = 32

14. 9 teaspoons = ___ tablespoons

Find the perimeter and area of each rectangle.

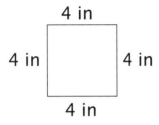 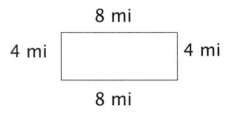

15. A = _____

16. P = _____

17. A = _____

18. P = _____

19. Dad gave Sandi $9 in bills for her quarter collection. How many quarters did Sandi give to Dad?

20. Sierra put five gallons of gasoline in her car and two gallons in a can for her lawn mower. How many quarts of gasoline did she buy altogether?

SYSTEMATIC REVIEW

16E

Multiply.

1. (6)(9) = _____

2. 7 × 4 = _____

3. 8 × 6 = _____

4. 10 · 5 = _____

5.   4
 × 3

6.   9
 × 7

7.   1
 × 0

8.   9
 × 4

9.   1
 × 1

10.   6
 × 7

11.   9
 × 9

12.   4
 × 8

Solve for the unknown and fill in the blanks.

13. 25 pennies = ___ nickels

14. 40 cents = ___ dimes

Find the perimeter and area of each rectangle.

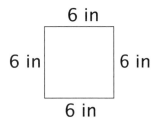
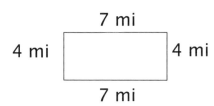

15. A = _____

16. P = _____

17. A = _____

18. P = _____

19. Sherri needed enough horseshoes for her six white horses and her two black horses. How many horseshoes did she need altogether?

20. Harriet decided to run two ads in the local newspaper. They each cost $4 per day. She ran them both for four days. How much did she have to pay? (First find out how much it cost to run one ad for four days.)

SYSTEMATIC REVIEW 16F

Multiply.

1. (7)(3) = _____

2. 8 × 9 = _____

3. 5 × 7 = _____

4. 6 · 9 = _____

5.   2
  × 8

6.   9
  × 3

7.   7
  × 2

8.   7
  × 4

9.   5
  × 9

10.   1 0
  × 7

11.   6
 × 4

12.   1
 × 5

Solve for the unknown and fill in the blanks.

13. 16 pints = ___ quarts

14. 27 feet = ___ yards

SYSTEMATIC REVIEW 16F

Find the perimeter of each square.

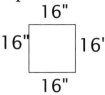

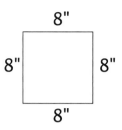

15. P = _____

16. P = _____

Find the area and perimeter of the rectangle.

```
        9 ft
      ┌──────┐
5 ft  │      │  5 ft
      └──────┘
        9 ft
```

17. A = _____

18. P = _____

19. Jamie counted the money in his pocket and found he had 50 cents. The money was all in dimes. How many dimes did Jamie have in his pocket?

20. Bonnie buttoned the jackets of her three boys and three girls. Each jacket had nine buttons. How many buttons did Bonnie button in all?

APPLICATION AND ENRICHMENT

16G

Multiply each number in the top row by four. Write the answers in the bottom row.

0	1	2	3	4	5	6	7	8	9	10
0	4									

Look at the numbers in the bottom row. Start at 0 on the octagon and connect the dots in order, using the numbers in the units place of each answer in the bottom row.

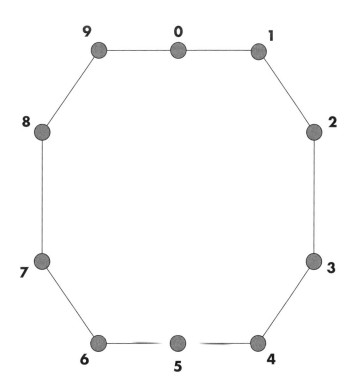

1. What shape did you make?

2. How many points does your shape have?

GAMMA APPLICATION AND ENRICHMENT 16G

Skip count by four. Start at the star and connect the dots all the way to 96. You may use the chart on 16A in the student book to help you if needed. Use the picture to practice skip counting by four.

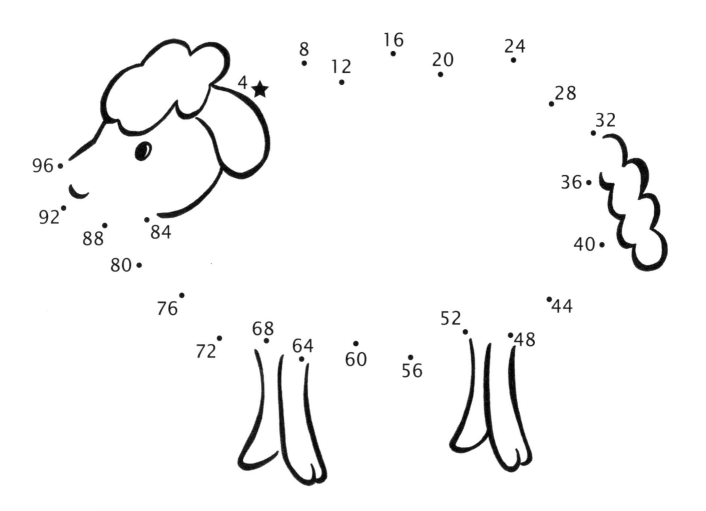

Six sick sheep strayed out of Sam's shed. Sam sadly spent $4 per sheep to shepherd the sheep to shelter.

How much did Sam pay altogether?

_____

LESSON PRACTICE 17A

Skip count by 7. Write the correct numbers in the squares with lines.

1.

						7
						___
						___
						___
						35
						___
						___
						___
						___
						70

Skip count and write the numbers.

2. 7, ___, ___, ___, ___, 42, ___, ___, ___, ___

LESSON PRACTICE 17A

Use "mittens" to help you multiply by multiples of 10. The first one has been done for you.

3. $\begin{array}{r} 3\,0 \\ \times\ 3 \end{array}$ → $\begin{array}{r} 3\ \\ \times\ 3 \\ \hline 9\ \end{array}$ → $\begin{array}{r} 3\,0 \\ \times\ 3 \\ \hline 9\,0 \end{array}$

4. $\begin{array}{r} 2\,0 \\ \times\ 2 \end{array}$

5. $\begin{array}{r} 4\,0 \\ \times\ 3 \end{array}$

Use skip counting to solve the word problems.

6. A father gave $7.00 to each of his five children. How much money did he give them altogether?

7. Kites cost seven dollars each. What is the cost of three kites?

8. Scott traveled for six hours at seven miles an hour. How far did he go?

LESSON PRACTICE 17B

Skip count by 7. Write the correct numbers in the squares with lines.

1.

						7
						__
						__
						28
						__
						__
						__
						__
						63
						__

Skip count and write the numbers.

2. ___, 14, ___, ___, ___, 42, ___, ___, ___, ___

LESSON PRACTICE 17B

Use "mittens" to help you multiply by multiples of 10.

3. 40
    × 5

4. 30
    × 6

5. 20
    × 3

Use skip counting to solve the word problems.

6. There are seven rows with eight apple trees in each row. How many trees are in the orchard?

7. Anna invited seven friends to her sleepover. Each friend ate two cookies. How many cookies were eaten altogether?

8. There are exactly four weeks in February. How many days are there in February? (seven days in a week) _____

LESSON PRACTICE 17C

Skip count by 7. Write the correct numbers in the squares with lines.

1.

							___
							14
							___
							___
							___
							___
							56
							___
							___

Skip count and write the numbers.

2. 7, ___, ___, 28, ___, ___, ___, ___, ___, ___

Use "mittens" to help you multiply by multiples of 10.

3. 2 0
   × 6
   ___

4. 5 0
   × 3
   ___

LESSON PRACTICE 17C

5. 8 0
   × 2
   ───

Use skip counting to solve the word problems.

6. Seven men each brought seven dogs to the dog show. How many dogs came to the show? _____

7. Nine monkeys each ate seven bananas. How many bananas were eaten in all?

8. New gloves cost $7 a pair. How much money is needed for six pairs of gloves?

SYSTEMATIC REVIEW

Skip count and write the numbers.

1. ___, ___, 21, ___, ___, ___, ___, ___, 63, ___

Fill in the blanks in the numerators and denominators to name the equivalent fractions.

2.

$$\frac{4}{7} = \frac{}{14} = \frac{}{} = \frac{16}{} = \frac{}{35}$$

Multiply.

3. (4)(9) = _____

4. 3×8 = _____

5. 2×1 = _____

6. $3 \cdot 4$ = _____

7. 7
   $\times$ 4

8. 8
   $\times$ 6

9. 4 0
   $\times$ 9

10. 5 0
    $\times$ 2

SYSTEMATIC REVIEW 17D

Solve for the unknown.

11. $3X = 18$

12. $8Y = 80$

13. $1R = 5$

14. $7Q = 63$

QUICK REVIEW

The symbols < and > are used to show inequalities. Read 2 < 3 as "Two is less than three," and read 4 > 1 as "Four is greater than one." The small end of the symbol always points to the lesser value. The open or larger end points to the greater value.

Write <, >, or = in the oval. The first one has been done for you.

15. $3 + 2$ ⬭<⬭ 3×2

16. $8 + 9$ ◯ $9 + 8$

17. 31 ◯ 13

18. Grandpa gave Riley $7 worth of quarters. How many quarters did Riley receive?

19. A rectangle is five inches long and three inches wide. What is its perimeter?

20. Tonya had 65 pennies, but she lost 36 of them. How many pennies does she have left?

SYSTEMATIC REVIEW 17E

Skip count and write the numbers.

1. ____, 12, ____, ____, 30, ____, ____, ____, ____, ____

Fill in the blanks in the numerators and denominators to name the equivalent fractions.

2.

$$\frac{6}{7} = \frac{}{} = \frac{}{21} = \frac{24}{} = \frac{}{}$$

Multiply.

3. (4)(6) = ____

4. 4 × 8 = ____

5. 6 × 3 = ____

6. 9 · 8 = ____

7.   4
   × 4

8.   8
   × 5

9.   6 0
   × 9

10. 3 0
   × 7

SYSTEMATIC REVIEW 17E

Solve for the unknown.

11. 6X = 54

12. 5Y = 10

13. 6R = 0

14. 2Q = 12

Write <, >, or = in the oval.

15. 9 qt ◯ 2 gal

16. 6 × 3 ◯ 2 × 9

17. 7 + 8 ◯ 20 − 2

18. During his first week of work, Adam spent 32 hours pulling weeds. The second week he did the same. The third week he had other things to do, so he spent only 17 hours pulling weeds. How many hours did he spend pulling weeds during those three weeks?

19. Another name for 10 years is a decade. How many years are in five decades?

20. Joel bought nine quarts of grape juice. Amy bought six quarts of apple juice. In all, how many pints of juice did they buy?

SYSTEMATIC REVIEW

17F

Skip count and write the numbers.

1. ____, ____, 21, ____, ____, ____, ____, ____, ____, 70

Fill in the blanks in the numerators and denominators to name the equivalent fractions.

2.

$\dfrac{4}{5} = \dfrac{}{} = \dfrac{}{15} = \dfrac{16}{} = \dfrac{}{}$

Multiply.

3. (3)(4) = _____

4. 9 × 9 = _____

5. 6 × 8 = _____

6. 4 · 3 = _____

7.   5
  × 9

8.   1 0
  × 7

9.   9 0
  × 3

10.   6 0
  × 4

SYSTEMATIC REVIEW 17F

Solve for the unknown.

11. $7X = 14$

12. $4Y = 36$

13. $6R = 42$

14. $5Q = 15$

Write <, >, or = in the oval.

15. $4 \times 9 \bigcirc 6 \times 5$

16. $5 + 6 \bigcirc 16 - 4$

17. 6 yd $\bigcirc$ 18 ft

18. Alyssa has five dimes and three nickels. Does she have enough money to buy a candy bar that costs 75 cents?

    What is the difference between the cost of the candy bar and the amount of money she has?

19. Derek drew eight separate squares. How many straight lines did he draw?

20. Mr. Smith has lived for nine decades. For how many years has he lived?

APPLICATION AND ENRICHMENT 17G

You can use an arrangement called an *array* to see how multiplication and addition are related. Look at the examples to see how this works.

These two arrays show ways to make nine.

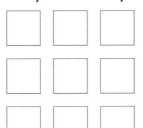
This shows 3 × 3 = 9
or 3 + 3 + 3 = 9.

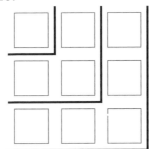
Look closely at the groups. This shows 1 + 3 + 5 = 9.

Here are three ways to make 16.

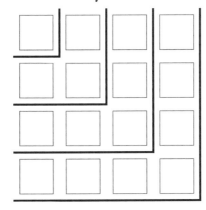

4 × 4 = 16
4 + 4 + 4 + 4 = 16
1 + 3 + 5 + 7 = 16

Find the answer three different ways.

___ × ___ = ___

___ + ___ + ___ + ___ + ___ = ___

___ + ___ + ___ + ___ + ___ = ___

APPLICATION AND ENRICHMENT 17G

Color the picture. Complete each step in the order given for best results. If you have already colored a number, do not color it again in the next step.

If you say the number when you skip count by 10 to 100, color the space yellow.
If you say the number when you skip count by 6 to 60, color the space pink.
If you say the number when you skip count by 3 to 30, color the space black.
If you say the number when you skip count by 4 to 40, color the space blue.

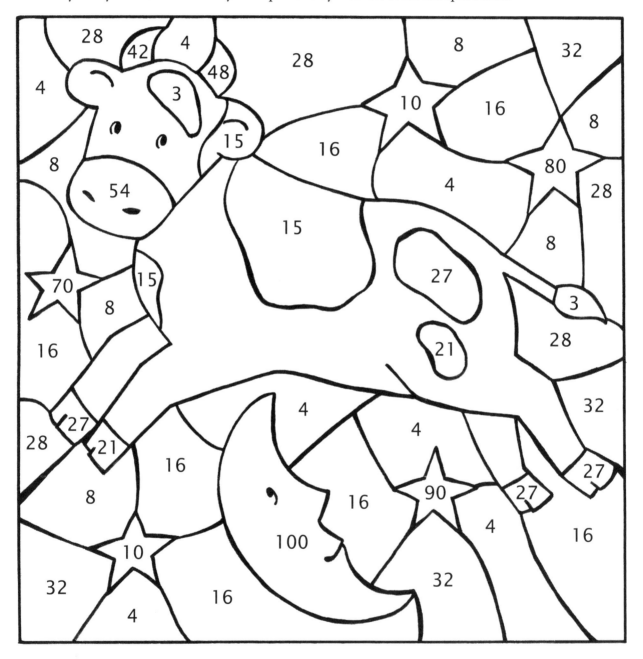

LESSON PRACTICE 18A

Find the answer by multiplying.

1. $7 \times 9 =$ _____

2. $7 \times 7 =$ _____

3. $10 \times 7 =$ _____

4. $7 \cdot 6 =$ _____

5. 7
   $\underline{\times\ 4}$

6. 7
   $\underline{\times\ 8}$

7. 7
   $\underline{\times\ 5}$

8. 3
   $\underline{\times\ 7}$

9. Seven counted eight times equals _____.

10. Seven counted seven times equals _____.

LESSON PRACTICE 18A

Color all the boxes that have a number you would say when skip counting by 7. Finish the pattern and use it to answer the problems beside the chart.

11.

0	1	2	3	4	5	6	7	8	9
10	11	12	13	14	15	16	17	18	19
20	21	22	23	24	25	26	27	28	29
30	31	32	33	34	35	36	37	38	39
40	41	42	43	44	45	46	47	48	49
50	51	52	53	54	55	56	57	58	59
60	61	62	63	64	65	66	67	68	69
70	71	72	73	74	75	76	77	78	79
80	81	82	83	84	85	86	87	88	89
90	91	92	93	94	95	96	97	98	99

$7 \times 11 =$ ___

$7 \times 12 =$ ___

$7 \times 13 =$ ___

Use "mittens" to help you multiply by multiples of 100. The first one has been done for you.

12. 300 →   3 🧤 →   300
    × 3   × 3   ×   3
    _____   _____   _____
            9 🧤   900

13. 200
    × 4
    _____

14. 100
    × 9
    _____

15. A seamstress sewed seven skirts a day. How many did she sew in seven days?

16. How many days are in six weeks?

LESSON PRACTICE 18B

Find the answer by multiplying.

1. 7 × 8 = _____
2. 7 × 4 = _____

3. 9 × 7 = _____
4. 7 · 7 = _____

5. 7
   × 3
   ‾‾‾

6. 7
   × 5
   ‾‾‾

7. 7
   × 8
   ‾‾‾

8. 6
   × 7
   ‾‾‾

Use "mittens" to help you multiply by multiples of 100.

9. 4 0 0
   ×   2
   ‾‾‾‾‾

10. 2 0 0
    ×   3
    ‾‾‾‾‾

11. 1 0 0
    ×   6
    ‾‾‾‾‾

12. 3 0 0
    ×   2
    ‾‾‾‾‾

13. Use your blocks to build the numbers you say when skip counting by 7.

LESSON PRACTICE 18B

Skip count and write the missing numbers. Then fill in the missing factors under the lines.

14.
$$\frac{0}{7\times 0} \quad \frac{}{7\times \underline{}} \quad \frac{}{7\times 2} \quad \frac{}{7\times \underline{}} \quad \frac{28}{7\times \underline{}} \quad \frac{}{7\times 5}$$

$$\frac{}{7\times \underline{}} \quad \frac{}{7\times \underline{}} \quad \frac{}{7\times 8} \quad \frac{}{7\times \underline{}} \quad \frac{70}{7\times \underline{}}$$

15. Seven counted seven times equals _____ .

16. Scott spent eight weeks at his grandparents' house. For how many days did he visit them?

17. A fortnight is two weeks. How many days are in a fortnight?

18. One hundred soldiers marched in each group. There were seven groups. How many soldiers marched in all?

LESSON PRACTICE

18C

Find the answer by multiplying.

1. $7 \times 7 =$ _____

2. $7 \times 2 =$ _____

3. $7 \times 8 =$ _____

4. $9 \cdot 7 =$ _____

5.
```
   5
 × 7
```

6.
```
   6
 × 7
```

7.
```
   7
 × 3
```

8.
```
   7
 × 4
```

Use "mittens" to help you multiply by multiples of 100.

9.
```
 200
 × 2
```

10.
```
 100
 × 5
```

11.
```
 300
 × 3
```

12.
```
 200
 × 4
```

13. Use your blocks to build the numbers you say when skip counting by 7.

LESSON PRACTICE 18C

Skip count and write the missing numbers. Then fill in the missing factors under the lines.

14.

$$\frac{0}{7\cdot 0} \quad \frac{}{7\cdot \underline{}} \quad \frac{14}{7\cdot \underline{}} \quad \frac{}{7\cdot 3} \quad \frac{}{7\cdot \underline{}} \quad \frac{}{7\cdot \underline{}}$$

$$\frac{}{7\cdot 6} \quad \frac{}{7\cdot \underline{}} \quad \frac{56}{7\cdot \underline{}} \quad \frac{}{7\cdot 9} \quad \frac{}{7\cdot \underline{}}$$

15. Seven counted one time equals _____ .

16. Mrs. Brown wants to put four slices of bologna in each sandwich. She plans to make seven sandwiches. How many slices of bologna does she need?

17. Sally went skating 10 times a month for 7 months. How many times did she go skating?

18. Four hundred blackbirds swooped over our backyard. How many wings were beating over the yard?

SYSTEMATIC REVIEW

18D

Multiply.

1. (6)(2) = _____

2. 7 × 8 = _____

3. 3 × 3 = _____

4. 7 · 2 = _____

5.   1
  × 9

6.   7
  × 7

7.   1 0
  × 8

8.   4
  × 3

9.   7 0
  × 6

10.   9
  × 7

11.   4 0
  × 7

12.   2 0 0
  ×   3

Solve for the unknown and fill in the blanks.

13. 27 feet = ___ yards

14. 15 teaspoons = ___ tablespoons

SYSTEMATIC REVIEW 18D

Write <, >, or = in the oval.

15. $5 \times 4 \bigcirc 3 \times 7$

16. $16 + 15 \bigcirc 35 - 6$

17. $5 \times 7 \bigcirc 7 \times 5$

18. Another name for 100 years is a century. Dad told Craig that an old house in their town is three centuries old. How many years old is the house?

19. A rectangular room measures seven feet by eight feet. What is the area of the room?

    What is its perimeter?

20. Karena counted the cars going past her house. The first day she counted 79, the second day she counted 82, and the third day she counted 113 cars. How many cars went past her house during those three days?

SYSTEMATIC REVIEW 18E

Multiply.

1. (9)(7) = _____
2. 2×5 = _____
3. 3×6 = _____
4. $7 \cdot 7$ = _____

5.   2
  × 9

6.   8
  × 7

7.   5
  × 6

8.   1 0
  × 4

9.   1 0
  × 7

10.   6
  × 7

11.   6 0
  × 6

12. 1 0 0
 ×   8

Solve for the unknown and fill in the blanks.

13. 14 pints = ___ quarts
14. 49 days = ___ weeks

SYSTEMATIC REVIEW 18E

Write <, >, or = in the oval.

15. 28 ◯ 4 × 7

16. 40 − 19 ◯ 12 + 10

17. 5 nickels ◯ 3 dimes

18. The newspaper had a story about a woman who was one century old. How many years old was she?

19. Five people have decided to work together to clean up 15 miles of road in their town. Each person has cleaned two miles of road. How many miles are left to clean?

20. Rod dug for fishing worms in a plot of ground that measured five feet by five feet. He was able to find one fishing worm per square foot. After his fishing trip, he had 11 worms left. How many worms did he use?

SYSTEMATIC REVIEW 18F

Multiply.

1. (3)(7) = _____

2. 10 × 7 = _____

3. 3 × 4 = _____

4. 8 · 7 = _____

5.   9
   × 9

6.   8
   × 4

7.   7
   × 7

8. 1 0
   × 2

9.   8
   × 9

10.   4
    × 7

11. 7 0
    × 5

12. 2 0 0
    ×   3

Solve for the unknown and fill in the blanks.

13. 36 quarts = ___ gallons

14. 24 quarters = $___

SYSTEMATIC REVIEW 18F

Write <, >, or = in the oval.

15. 6 × 7 ◯ 10 × 4

16. 4 × 90 ◯ 300 + 60

17. 8 + 8 ◯ 30 − 14

18. For a science project, Cody recorded the temperature every morning for nine weeks. How many daily temperatures did he record for his project?

19. Sue bought a hat for each of her three sons and her five daughters. The hats cost $3 apiece. How much did she spend in all?

20. Every morning, Adam rode his bike three miles to work. At the end of the day, he took a longer, five-mile route home. How many miles did Adam ride in five days?

APPLICATION AND ENRICHMENT

18G

Multiply each number in the top row by seven. Write the answers in the bottom row.

0	1	2	3	4	5	6	7	8	9	10
0	7									

Look at the numbers in the bottom row. Start at 0 on the decagon (ten-sided shape) and connect the dots in order, using the numbers in the units place of each answer in the bottom row.

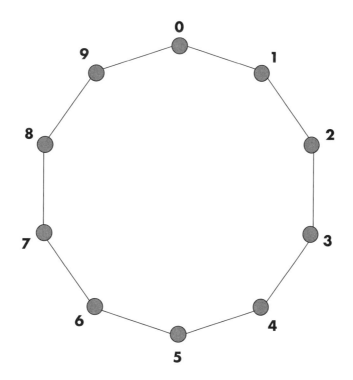

1. What shape did you make?

2. How many points does your shape have?

GAMMA   APPLICATION AND ENRICHMENT 18G

APPLICATION AND ENRICHMENT 18G

Skip count by seven. Start at the star and connect the dots all the way to 98. You may use the chart on 18A in the student book to help you if needed. Use the picture to practice skip counting by seven.

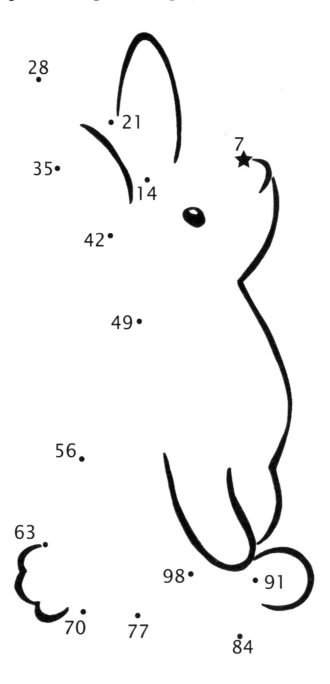

I saw Bobby Bunny bounce across my yard every day for eight weeks. For how many days did I watch him?

_____

LESSON PRACTICE 19A

Skip count by 8. Write the correct numbers in the squares with lines.

1.

								8
								__
								__
								32
								__
								__
								__
								__
								80

Skip count and write the numbers.

2. 8, ___, ___, ___, ___, 48, ___, ___, ___, ___

Find the missing multiples of 3 and 8 in the equivalent fractions.

3. $\dfrac{3}{8} = \dfrac{}{} = \dfrac{9}{} = \dfrac{}{32} = \dfrac{}{} = \dfrac{}{56} = \dfrac{24}{} = \dfrac{}{} = \dfrac{}{}$

LESSON PRACTICE 19A

Use skip counting to solve the problems.

4. How many sides are on the stop signs?

5. The houses on the street all have the same number of windows. Eight homes have a total of 56 windows. How many windows does each home have?

6. Each bridesmaid's dress required eight yards of fabric. How many yards of fabric must be bought to make dresses for a total of six bridesmaids?

7. There are eight pints in a gallon. How many pints are in eight gallons?

8. A spider has eight legs. There are four spiders on my wall. How many spider legs are on my wall?

LESSON PRACTICE 19B

Skip count by 8. Write the correct numbers in the squares with lines.

1.

							___
							16
							___
							___
							40
							___
							___
							___
							___

Skip count and write the numbers.

2. ____, ____, 24, ____, ____, ____, ____, ____, 72, ____

Find the missing multiples of 6 and 8 in the equivalent fractions.

3. $\dfrac{6}{8} = \dfrac{}{} = \dfrac{}{} = \dfrac{24}{} = \dfrac{}{48} = \dfrac{}{} = \dfrac{}{} = \dfrac{}{80}$

LESSON PRACTICE 19B

Use skip counting to solve the problems.

4. How many pints are in six gallons?

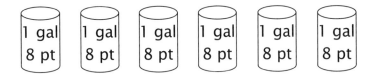

5. A stop sign has eight sides. Cody drew three stop signs on a piece of paper. How many sides did he draw?

6. Jack makes $8 an hour gardening. He worked eight hours. How much money did he earn?

7. Lyle went to Vermont and bought two gallons of maple syrup in pint jugs. How many jugs of syrup did Lyle buy?

8. How many legs are on nine eight-legged scorpions?

LESSON PRACTICE

19C

Skip count by 8. Write the correct numbers in the squares with lines.

1.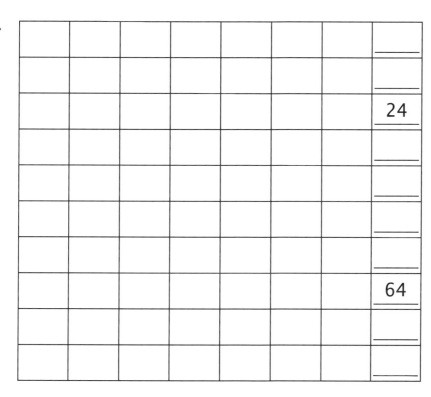

Skip count and write the numbers.

2. ___, ___, ___, ___, ___, ___, ___, ___, ___, 80

Find the missing multiples of 4 and 8 in the equivalent fractions.

3. $\dfrac{4}{8} = \dfrac{}{} = \dfrac{}{} = \dfrac{16}{} = \dfrac{}{} = \dfrac{}{} = \dfrac{}{} = \dfrac{}{12} = \dfrac{}{}$

LESSON PRACTICE 19C

Use skip counting to solve the problems.

4. How many legs are on three spiders?

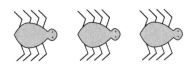

5. Fred the Fearless Diver encountered a family of five octopuses. Octopuses have eight arms each. How many arms were waving at Fred?

6. Four more octopuses swam up to Fred (#5). What is the total number of arms now surrounding him?

7. Sarah worked eight hours a week. When she had worked a total of 80 hours, how many weeks had she worked?

8. Brandon counted seven stop signs on the way to work. How many sides did he count in all?

SYSTEMATIC REVIEW

19D

Skip count and write the numbers.

1. ____, 16, ____, ____, ____, ____, ____, ____, ____, 80

Fill in the blanks in the numerators and denominators to name the equivalent fractions.

2.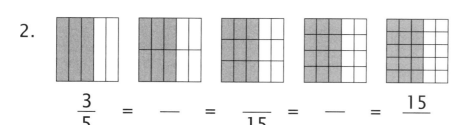

$$\frac{3}{5} = \frac{}{} = \frac{}{15} = \frac{}{} = \frac{15}{}$$

Multiply.

3. (8)(3) =

4. 5 × 3 =

5. 7 × 7 =

6. 2 · 4 =

7.   7
  × 8

8.   3
  × 6

9.   4 0
  × 8

10.   2 0 0
  ×   2

Solve for the unknown.

11. 9X = 54

12. 4Y = 16

SYSTEMATIC REVIEW 19D

13. 8R = 0

14. 7Q = 35

QUICK REVIEW

Add three-digit numbers just as you do two-digit numbers. Your answer may include the thousands place.

Add. The first one has been done for you.

15.
```
    1
   3 4 5
 + 7 2 9
 -------
 1,0 7 4
```

16.
```
   1 2 8
 + 6 3 5
 -------
```

17.
```
   2 1 2
 + 8 7 2
 -------
```

18. Square dance circles are made up of four couples. How many square dancers are in six circles? (Be careful!)

19. Max's room is three yards wide. Each of his shoes is exactly one foot long. How many of his shoes would he need to arrange end to end to reach across his room?

20. Craig led the car race for 179 laps at the beginning and then fell behind. He took the lead again for another 143 laps and won the race. For how many laps altogether did he lead?

SYSTEMATIC REVIEW

19E

Skip count and write the numbers.

1. 7, ___, ___, ___, ___, ___, ___, ___, ___, ___

Fill in the blanks in the numerators and denominators to name the equivalent fractions.

2.

$\frac{2}{6} = \frac{}{} = \frac{6}{} = \frac{}{24} = \frac{}{}$

Multiply.

3. (8)(2) = _____

4. 5 × 8 = _____

5. 9 × 8 = _____

6. 3 · 6 = _____

7. 7
   × 7
   ___

8. 1 0
   × 5
   ___

9. 6 0
   × 7
   ___

10. 1 0 0
    ×   9
    ___

GAMMA SYSTEMATIC REVIEW 19E

SYSTEMATIC REVIEW 19E

Solve for the unknown.

11. 3X = 27

12. 6Y = 36

13. 5R = 5

14. 4Q = 20

Add.

15. 6 0 1
   + 5 1 3

16. 2 4 5
   + 1 8 9

17. 5 3 8
   + 2 5 1

18. Over the course of the summer, Eddie ate three gallons of vanilla ice cream and four gallons of chocolate ice cream. How many pints did he eat?

19. Rover, the poodle, weighed two pounds. Fido, the cat, weighed four times as much as Rover. How much did Rover and Fido weigh when they stood on the scale together?

20. Claire loved to collect coins. She collected 74 the first year and 98 the second year. The third year she managed to collect 206 more coins. How many coins did she have after three years?

SYSTEMATIC REVIEW 19F

Skip count and write the numbers.

1. ____, ____, ____, ____, 40, ____, ____, ____, 72, ____

Fill in the blanks in the numerators and denominators to name the equivalent fractions.

2. $\frac{1}{7} = \frac{}{} = \frac{}{21} = \frac{4}{} = \frac{}{}$

Multiply.

3. (8)(10) = ____

4. 6 × 8 = ____

5. 8 × 4 = ____

6. 3 · 10 = ____

7. 1
　× 1

8. 8 0
　× 4

9. 1 0
　× 5

10. 1 0 0
　×　 2

SYSTEMATIC REVIEW 19F

Solve for the unknown.

11. $8X = 24$

12. $7Y = 49$

13. $4R = 12$

14. $7Q = 14$

Add.

15. 452
    +318

16. 711
    +206

17. 153
    +592

18. Hans went to the doctor, who charged him $45 for his visit. Hans had only $27 with him. How much more does he still owe the doctor?

19. Roy bought six cans of gasoline. Each can held five gallons. When Roy got home, he put seven gallons of gasoline in his tractor. How much gasoline was left?

20. Doug's storage room measures 5 feet wide and 10 feet long. What is the area of the floor?

Doug has 35 floor tiles that each measure one foot square. How many more tiles does he need to buy to cover the floor of his storage room?

APPLICATION AND ENRICHMENT 19G

Some creatures have eight arms or legs. Skip count each group to see how many arms or legs altogether.

1. Six scorpions have _____ legs.

2. Nine spiders have _____ legs.

3. Eight ticks have _____ legs.

4. Seven octopuses have _____ arms.

APPLICATION AND ENRICHMENT 19G

Kate boiled down maple sap to make maple syrup. She made a chart to show how much syrup she made each day.

Finish filling in the chart and use it to answer the questions.

Day	Number of Gallons	Number of Pints
Sunday	0	0
Monday	3	24
Tuesday	4	
Wednesday	6	
Thursday	2	
Friday	7	
Saturday	5	

1. On which day did Kate make the most gallons of maple syrup?

2. On which day did Kate make the fewest gallons of maple syrup?

3. How many pints of syrup did she make on Tuesday?

4. How many pints of syrup did she make on Friday?

5. How many gallons of syrup did she make altogether that week?

6. How many pints of syrup did she make altogether that week?

LESSON PRACTICE 20A

Find the answer by multiplying.

1. $8 \times 9 =$ _____
2. $7 \times 8 =$ _____

3. $6 \cdot 8 =$ _____
4. $8 \cdot 10 =$ _____

5. $5 \times 8 =$ _____
6. $1 \times 8 =$ _____

7. $(8)(3) =$ _____
8. $(8)(2) =$ _____

9. 8
   $\times\,8$

10. 8
    $\times\,9$

11. 4
    $\times\,8$

12. 8
    $\times\,7$

13. Eight counted zero times equals _____.

14. Eight counted eight times equals _____.

LESSON PRACTICE 20A

Color all the boxes that have a number you would say when skip counting by 8. Finish the pattern. Compare it with the chart for counting by 4.

15.

0	1	2	3	4	5	6	7	8	9
10	11	12	13	14	15	16	17	18	19
20	21	22	23	24	25	26	27	28	29
30	31	32	33	34	35	36	37	38	39
40	41	42	43	44	45	46	47	48	49
50	51	52	53	54	55	56	57	58	59
60	61	62	63	64	65	66	67	68	69
70	71	72	73	74	75	76	77	78	79
80	81	82	83	84	85	86	87	88	89
90	91	92	93	94	95	96	97	98	99

Multiply by 8 to find the number of sides on the stop signs.

16.

$8 \times 10 =$ _____

17. A shape with eight sides like a stop sign is called an octagon. How many sides are there on six octagons?

18. I can buy eight oranges for a dollar. How many oranges can I buy with three dollars?

LESSON PRACTICE 20B

Find the answer by multiplying.

1. $8 \times 4 =$ _____
2. $8 \times 8 =$ _____

3. $5 \cdot 8 =$ _____
4. $7 \cdot 8 =$ _____

5. $9 \times 8 =$ _____
6. $10 \times 8 =$ _____

7. $(8)(6) =$ _____
8. $(8)(3) =$ _____

9.   9
   × 8

10.   8
   × 2

11.   8
   × 8

12.   8
   × 1

13. Use your blocks to build the numbers you say when skip counting by 8.

LESSON PRACTICE 20B

Skip count and write the missing numbers. Then fill in the missing factors under the lines.

14.

$$\frac{0}{(8)(0)} \quad \frac{}{(8)(\_)} \quad \frac{}{(8)(\_)} \quad \frac{}{(8)(\_)} \quad \frac{}{(8)(\_)} \quad \frac{}{(8)(5)}$$

$$\frac{}{(8)(\_)} \quad \frac{}{(8)(\_)} \quad \frac{64}{(8)(\_)} \quad \frac{}{(8)(\_)} \quad \frac{}{(8)(10)}$$

Multiply by 8 to find the number of sides on the stop signs.

15.

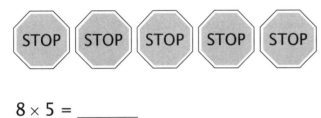

8 × 5 = _____

16. Eight counted four times equals _____.

17. For a family reunion, we ordered 7 pizzas with 8 slices in each pizza. How many slices did we receive in all?

18. Nine gallons of punch were made for the reunion. If one pint is enough for one person, how many people can be served?

LESSON PRACTICE 20C

Find the answer by multiplying.

1. 6 × 8 = _____
2. 8 × 2 = _____

3. 8 · 8 = _____
4. 8 · 9 = _____

5. 8 × 3 = _____
6. 5 × 8 = _____

7. (7)(8) = _____
8. (8)(10) = _____

9.   4
  × 8

10.   8
  × 1

11.   6
  × 8

12.   8
  × 9

13. Eight counted eight times equals _____

14. Eight counted five times equals _____

LESSON PRACTICE 20C

Color all the boxes that have a number you would say when skip counting by 8. Finish the pattern and use it to answer the problems beside the chart.

15.

0	1	2	3	4	5	6	7	8	9
10	11	12	13	14	15	16	17	18	19
20	21	22	23	24	25	26	27	28	29
30	31	32	33	34	35	36	37	38	39
40	41	42	43	44	45	46	47	48	49
50	51	52	53	54	55	56	57	58	59
60	61	62	63	64	65	66	67	68	69
70	71	72	73	74	75	76	77	78	79
80	81	82	83	84	85	86	87	88	89
90	91	92	93	94	95	96	97	98	99

$8 \times 11 = $ ____

$8 \times 12 = $ ____

Multiply by 8 to find the number of sides on all the stop signs.

16.

$8 \times 7 = $ _____

17. How many sides are on four octagons?

18. Some large cars have eight-cylinder engines. How many cylinders would be in eight of those cars?

SYSTEMATIC REVIEW

20D

Multiply.

1. (7)(7) = _____

2. 8 × 6 = _____

3. 9 × 9 = _____

4. 3 · 1 = _____

5.   8
  × 7

6.   9
  × 3

7.   7
  × 6

8.   9
  × 5

9.   8 0
 × 8

10.   2
  × 0

11.   4 0
  × 4

12. 1 0 0
 ×   3

Write <, >, or = in the oval.

13. 7 × 7 ◯ 7 + 7

14. 32 + 9 ◯ 17 + 24

SYSTEMATIC REVIEW 20D

15. 3 gal ◯ 28 pt

QUICK REVIEW

When subtracting three-digit numbers, first regroup the tens if needed and then regroup from the hundreds to the tens place, if necessary. If the number in the tens place is zero, regroup and make it 10 before taking one from it to the units place.

Subtract. The first two have been done for you.

16. ³4̸ ⁱ⁴5̸ ¹1
    − 2 9 8
    ─────────
       1 5 3

17. ⁶7̸ ⁹1̸0̸ ¹3
    − 4 1 8
    ─────────
       2 8 5

18.   5 6 2
    − 3 7 4
    ─────────

19. Haley has 3 cute little spiders and 4 big hairy spiders as pets. Since spiders have 8 legs, how many legs do Haley's spiders have in all?

20. Nigel beat Richard in a game by a score of 458 to 328. What is the difference in their scores?

SYSTEMATIC REVIEW

20E

Multiply.

1. (8)(8) = _____

2. 7×7 = _____

3. 2×6 = _____

4. $5 \cdot 4$ = _____

5.   9
  × 4
  ―――

6.   6
  × 6
  ―――

7.   7
  × 8
  ―――

8.   3
  × 7
  ―――

9.   40
  × 3
  ―――

10.   2
  × 8
  ―――

11.   80
  × 3
  ―――

12. 200
  × 2
  ―――

SYSTEMATIC REVIEW 20E

Write <, >, or = in the oval.

13. 45 cents ◯ 8 nickels   14. 3 × 8 ◯ 6 × 4

15. 16 + 16 ◯ 42 − 9

Subtract.

16.   6 0 3
  − 1 1 8

17.   3 4 5
  − 1 4 2

18.   8 3 7
  − 1 0 8

19. John had eight bicycles and eight cars at his repair shop. Every single tire mounted on each car and each bicycle was flat. How many tires did John need to repair?

20. Kurt bought 244 pencils on sale. He gave 188 of them to his sister. How many pencils did he have left?

SYSTEMATIC REVIEW

Multiply.

1. (3)(6) = _____   2. 9 × 7 = _____

3. 8 × 4 = _____   4. 3 · 9 = _____

5. 8   6. 5
   × 8   × 0

7. 3   8. 7
   × 3   × 7

9. 70   10. 9
   × 4   × 9

11. 60   12. 400
    × 3   ×   2

SYSTEMATIC REVIEW 20F

Write <, >, or = in the oval.

13. (0)(9) ◯ 4 + 5   14. 19 + 28 ◯ 35 − 17

15. 16 qts ◯ 5 gal

Subtract.

16.   9 0 0
    − 1 2 3

17.   6 8 3
    − 2 5 4

18.   5 0 6
    − 3 4 4

19. Kent is shopping for two used lawn mowers for his golf course. He found a riding mower for $549 and a push mower for $86. He has $350 to use. How much more money does Kent need to buy the two mowers?

20. Marlene counted the wings on the geese that were flying overhead. There were 18 wings. How many geese did she see?

APPLICATION AND ENRICHMENT

20G

Multiply each number in the top row by eight. Write the answers in the bottom row.

0	1	2	3	4	5	6	7	8	9	10
0	8									

Look at the numbers in the bottom row. Start at 0 on the circle and connect the dots in order, using the numbers in the units place of each answer in the bottom row.

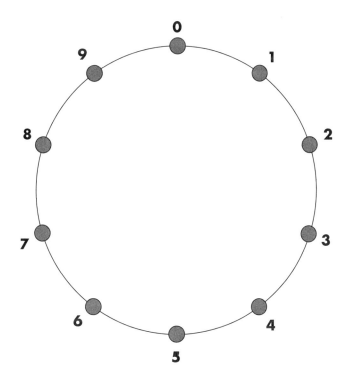

1. How many sides does your shape have?

2. Write the name of the shape, if you know it.

APPLICATION AND ENRICHMENT 20G

Skip count by eight. Start at the star and connect the dots all the way to 96. You may use the chart on 20A in the student book to help you if needed. Use the picture to practice skip counting by eight.

Ella Elephant remembered eight important things every evening. How many things did Ella remember in eight evenings?

_____

LESSON PRACTICE 21A

Multiply using standard notation and place-value notation. The first one has been done for you.

1. $\begin{array}{r} 21 \rightarrow \\ \times\,4 \uparrow \\ \hline 84 \end{array}$ $\begin{array}{r} 20+1 \\ \times\quad\ 4 \\ \hline 80+4 \end{array}$

2. $\begin{array}{r} 13 \rightarrow \\ \times\,2 \\ \hline \end{array}$ $\begin{array}{r} 10+3 \\ \times\quad\ 2 \\ \hline \end{array}$

3. $\begin{array}{r} 11 \rightarrow \\ \times\,7 \uparrow \\ \hline \end{array}$ $\begin{array}{r} 10+1 \\ \times\quad\ 7 \\ \hline \end{array}$

4. $\begin{array}{r} 21 \rightarrow \\ \times\,2 \uparrow \\ \hline \end{array}$ $\begin{array}{r} 20+1 \\ \times\quad\ 2 \\ \hline \end{array}$

5. $\begin{array}{r} 32 \rightarrow \\ \times\,3 \uparrow \\ \hline \end{array}$ $\begin{array}{r} 30+2 \\ \times\quad\ 3 \\ \hline \end{array}$

6. $\begin{array}{r} 14 \rightarrow \\ \times\,2 \uparrow \\ \hline \end{array}$ $\begin{array}{r} 10+4 \\ \times\quad\ 2 \\ \hline \end{array}$

7. $\begin{array}{r} 11 \rightarrow \\ \times\,9 \uparrow \\ \hline \end{array}$ $\begin{array}{r} 10+1 \\ \times\quad\ 9 \\ \hline \end{array}$

8. $\begin{array}{r} 24 \rightarrow \\ \times\,2 \uparrow \\ \hline \end{array}$ $\begin{array}{r} 20+4 \\ \times\quad\ 2 \\ \hline \end{array}$

LESSON PRACTICE 21A

9. $\begin{array}{r}123 \rightarrow \\ \times 2 \uparrow \\ \hline \end{array}$ $\begin{array}{r}100+20+3 \\ \times 2 \\ \hline \end{array}$

10. $\begin{array}{r}222 \rightarrow \\ \times 4 \uparrow \\ \hline \end{array}$ $\begin{array}{r}200+20+2 \\ \times 4 \\ \hline \end{array}$

11. Andy gave each of his 12 friends four books. How many books did Andy give away?

12. Three hundred ten cows waited in the barn. The veterinarian must give each of them three shots. How many shots must he give to finish the job?

LESSON PRACTICE 21B

Multiply using standard notation and place-value notation.

1. 21 →   20 + 1
    × 3 ↑   ×   3

2. 24 →   20 + 4
    × 2 ↑   ×   2

3. 22 →   20 + 2
    × 4 ↑   ×   4

4. 11 →   10 + 1
    × 6 ↑   ×   6

5. 14 →   10 + 4
    × 2 ↑   ×   2

6. 33 →   30 + 3
    × 3 ↑   ×   3

7. 110 →   100 + 10 + 0
    × 5 ↑ ×   5

LESSON PRACTICE 21B

8. $\begin{array}{r} 231 \rightarrow \\ \times 3 \uparrow \\ \hline \end{array}$ $\begin{array}{r} 200+30+1 \\ \times 3 \\ \hline \end{array}$

9. $\begin{array}{r} 424 \rightarrow \\ \times 2 \uparrow \\ \hline \end{array}$ $\begin{array}{r} 400+20+4 \\ \times 2 \\ \hline \end{array}$

10. $\begin{array}{r} 121 \rightarrow \\ \times 4 \uparrow \\ \hline \end{array}$ $\begin{array}{r} 100+20+1 \\ \times 4 \\ \hline \end{array}$

11. A certain rectangle is 12 inches long and two inches high. What is its area?

12. Two hundred fourteen soldiers marched by. How many marching legs passed us?

LESSON PRACTICE 21C

Multiply using standard notation and place-value notation.

1. 43 →   40 + 3
   ×2 ↑   ×   2

2. 32 →   30 + 2
   ×2 ↑   ×   2

3. 12 →   10 + 2
   ×3 ↑   ×   3

4. 11 →   10 + 1
   ×4 ↑   ×   4

5. 42 →   40 + 2
   ×2 ↑   ×   2

6. 31 →   30 + 1
   ×3 ↑   ×   3

7. 413 →   400 + 10 + 3
   × 2 ↑   ×   2

LESSON PRACTICE 21C

8. $\begin{array}{r}111 \rightarrow \\ \times 6 \uparrow \\ \hline \end{array}$ $\begin{array}{r}100+10+1 \\ \times 6 \\ \hline \end{array}$

9. $\begin{array}{r}103 \rightarrow \\ \times 3 \uparrow \\ \hline \end{array}$ $\begin{array}{r}100+00+3 \\ \times 3 \\ \hline \end{array}$

10. $\begin{array}{r}212 \rightarrow \\ \times 4 \uparrow \\ \hline \end{array}$ $\begin{array}{r}200+10+2 \\ \times 4 \\ \hline \end{array}$

11. Vincent planted four rows of tomatoes in his garden. Each row had 21 tomato plants. How many tomato plants did Vincent have in all?

12. Sean found a spider's egg case in his garden. As he watched, 111 little spiderlings hatched out. How many little spider legs came out of the egg case?

SYSTEMATIC REVIEW

21D

Multiply using standard notation and place-value notation.

1.   1 1   1 0 + 1
       × 5   ×   5

2.   1 2   1 0 + 2
       × 3   ×   3

3.   3 2 4   3 0 0 + 2 0 + 4
       × 2   ×   2

4.   3 2 2   3 0 0 + 2 0 + 2
       × 3   ×   3

Multiply.

5. (8)(7) = _____

6. 8 × 8 = _____

7. 7 × 7 = _____

8. 4 · 6 = _____

Challenge: Write the correct sign (+, −, ×) in each blank. The first one has been done for you.

9. 8 _×_ 5 = 40

10. 8 __ 5 = 13

11. 7 __ 6 = 1

12. 7 __ 6 = 42

SYSTEMATIC REVIEW 21D

Add or subtract.

13. 23
     45
    +17
    ———

14. 39
     24
    +88
    ———

15. 452
    -129
    ————

16. 283
    -216
    ————

17. What are the perimeter and the area of a room that measures 8 feet by 11 feet?

    P = _____, A = _____

18. Michelle made 9 astonishing discoveries every day for a week. The next day, she made 12 astonishing discoveries. How many discoveries did she make altogether in that time?

19. This morning Tony read 18 pages in one book and 52 pages in another book. How many pages did he read in all?

20. We purchased 150 logs. We used 123 of them to build our log cabin. How many logs were left over?

SYSTEMATIC REVIEW

21E

Multiply using regular notation and place-value notation.

1.
```
  1 2      1 0 + 2
× 4     ×       4
```

2.
```
  3 2      3 0 + 2
× 3     ×       3
```

3.
```
  2 2 1    2 0 0 + 2 0 + 1
×     4  ×               4
```

4.
```
  3 1 3    3 0 0 + 1 0 + 3
×     2  ×               2
```

Multiply.

5. (8)(4) = ___

6. 6 × 8 = ___

7. 7 × 9 = ___

8. 9 · 4 = ___

Write the correct sign (+, −, ×) in each blank.

9. 9 __ 9 = 18

10. 9 __ 9 = 81

11. 9 __ 8 = 72

12. 12 __ 4 = 8

SYSTEMATIC REVIEW 21E

Add or subtract.

13. 91
     25
    +42

14. 67
     13
    +50

15. 893
    -615

16. 314
    - 92

17. What are the perimeter and the area of a room that measures 6 feet by 7 feet?

    P = _____, A = _____

18. Two hundred fish live in each tank at the aquarium. How many fish are in three tanks?

19. Jamie brought three boxes of muffins to school. If each box contained 12 muffins, how many muffins did Jamie bring to school?

    John brought the same number of muffins as Jamie. How many muffins were brought to school in all?

20. I did 17 pushups on Monday, 22 on Tuesday, 24 on Wednesday, and 31 on Friday. How many pushups did I do in all that week?

SYSTEMATIC REVIEW

Multiply using standard notation and place-value notation.

1. 21   20 + 1
   × 2   ×   2

2. 11   10 + 1
   × 6   ×   6

3. 202   200 + 00 + 2
   × 3   ×   3

4. 444   400 + 40 + 4
   × 2   ×   2

Multiply.

5. (8)(6) = ___

6. 3 × 7 = ___

7. 6 × 10 = ___

8. 9 · 7 = ___

Write the correct sign (+, −, ×) in each blank.

9. 10 __ 3 = 7

10. 3 __ 4 = 12

11. 6 __ 3 = 9

12. 6 __ 3 = 18

Add or subtract.

13. 82
     13
   + 56

14. 71
     64
   + 26

15. 561
   - 19

16. 437
   - 365

17. How many feet of fence are needed to go around a yard that is 15 feet long and 12 feet wide?

18. Four hundred thirty-one birds flew over each night on their way south. How many birds flew over in two nights?

19. Paul went fishing one day and caught seven catfish and two trout every hour. If he fished for six hours, how many fish did he catch in all?

20. Johnny collected 11 eggs a day for seven days. How many eggs did he collect?

    Mom used 18 of the eggs to make angel food cake. How many eggs were left over?

APPLICATION AND ENRICHMENT

21G

If we build 2 × 23 with the blocks, it looks like this.

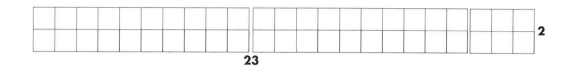

It is easier to multiply if we separate the blocks as shown below. Now we have two rectangles that are 2 by 20 and 2 by 3, but we still have the same number of blocks.

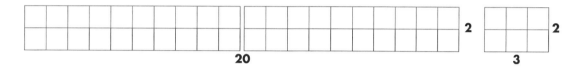

We can find the area of each rectangle and add the sums as shown.
2 × 23 = 2(20 + 3) = 2(20) + 2(3) = 40 + 6 = 46

Build each problem and then rearrange the blocks to match the numbers in the second line. Fill in the blanks to find the answer.

1. 2 × 13

   2(10 + 3) = 2(10) + 2(3) = ____ + ____ = ____

2. 3 × 21

   3(20 + 1) = 3(20) + 3(1) – ____ + ____ = ____

3. 2 × 32

   2(30 + 2) = 2(30) + 2(2) = ____ + ____ = ____

APPLICATION AND ENRICHMENT 21G

Find the answer three different ways. If you don't remember how to work with arrays, look at the application and enrichment page for lesson 17.

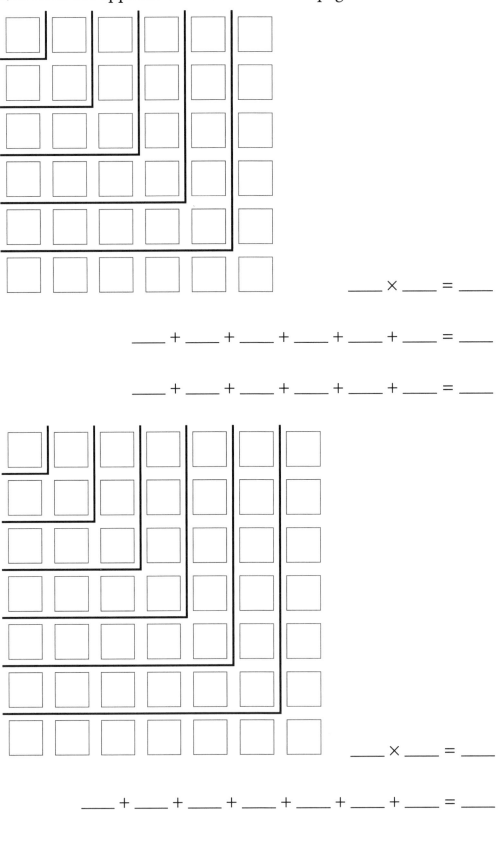

___ × ___ = ___

___ + ___ + ___ + ___ + ___ + ___ = ___

___ + ___ + ___ + ___ + ___ + ___ = ___

___ × ___ = ___

___ + ___ + ___ + ___ + ___ + ___ + ___ = ___

___ + ___ + ___ + ___ + ___ + ___ + ___ = ___

LESSON PRACTICE 22A

Round to the nearest 10.

1. 39 _____
2. 12 _____

3. 63 _____
4. 57 _____

5. 44 _____
6. 58 _____

Round to the nearest 100.

7. 361 _____
8. 209 _____

9. 682 _____

Round to the nearest 1,000.

10. 1,734 _____
11. 2,154 _____

12. 7,128 _____

Round the top number to the nearest 10 and estimate the answer. The first one has been done for you.

13. $\begin{array}{r} 25 \\ \times\ 2 \\ \end{array}$ → $\begin{array}{r} (30) \\ \times\ 2 \\ \hline (60) \end{array}$

14. $\begin{array}{r} 32 \\ \times\ 6 \\ \end{array}$ →

LESSON PRACTICE 22A

15. 19 →
     $\times$ 3

Round the top number to the nearest 100 and estimate the answer. The first one has been done for you.

16. 4 1 8 → (4 0 0)
     $\times$ 2 $\times$ 2
      (8 0 0)

17. 3 5 1 →
     $\times$ 3

18. 1 0 7 →
     $\times$ 4

19. Tom counted five rows of cars in the parking lot. Estimate how many cars in all if there are 26 cars in a row.

20. Ashley earned $35 a week babysitting. Estimate how much she earned in three weeks.

LESSON PRACTICE 22B

Round to the nearest 10.

1. 27 _____
2. 71 _____
3. 46 _____

Round to the nearest 100.

4. 519 _____
5. 734 _____
6. 154 _____
7. 856 _____
8. 128 _____
9. 173 _____

Round to the nearest 1,000.

10. 6,470 _____
11. 8,721 _____
12. 3,509 _____

Round the top number to the nearest 10 and estimate the answer.

13. 13 →
    × 2

14. 25 →
    × 2

LESSON PRACTICE 22B

15.   1 6 →
    × 7
    ─────

Round the top number to the nearest 100 and estimate the answer.

16.   3 7 5 →
    ×   3
    ───────

17.   1 2 9 →
    ×   4
    ───────

18.   6 1 1 →
    ×   2
    ───────

19. A round trip to Grandma's house is 24 miles. Estimate how many miles Jill traveled if she went to Grandma's house and back five times.

20. Molly liked a dress that cost $49. Estimate how much she spent if she bought three of those dresses in different colors.

LESSON PRACTICE 22C

Round to the nearest 10.

1. 85 _____
2. 38 _____
3. 21 _____

Round to the nearest 100.

4. 233 _____
5. 461 _____
6. 919 _____
7. 406 _____
8. 195 _____
9. 550 _____

Round to the nearest 1,000.

10. 7,553 _____
11. 1,390 _____
12. 2,816 _____

Round the top number to the nearest 10 and estimate the answer.

13. 48 →
     × 2

14. 36 →
     × 3

LESSON PRACTICE 22C

15. 23 →
    × 5
    ———

Round the top number to the nearest 100 and estimate the answer.

16. 1 0 6 →
     ×   8
     ————

17. 2 9 0 →
     ×   2
     ————

18. 3 5 1 →
     ×   6
     ————

19. A librarian discovered that she could fit 68 books on a shelf. Estimate how many books she could fit on nine shelves of the same length.

20. Mr. Brown rented a van that could carry 15 people. Estimate how many people could be carried in four vans.

SYSTEMATIC REVIEW

Round to the nearest 10.

1. 69 ____

2. 13 ____

3. 47 ____

Round the top number to the nearest 10 and estimate the answer.

4. 12 →
    × 8

5. 81 →
    × 9

6. 63 →
    × 5

Round the top number to the nearest 100 and estimate the answer.

7. 534 →
    ×   3

8. 351 →
    ×   6

9. 850 →
    ×   2

SYSTEMATIC REVIEW 22D

Multiply using standard notation and place-value notation.

10. 1 1   1 0 + 1   11. 2 1   2 0 + 1
     × 6   ×   6   × 4   ×   4

12. 1 1 3   1 0 0 + 1 0 + 3   13. 4 2 3   4 0 0 + 2 0 + 3
     ×   3   ×   3   ×   2   ×   2

Write <, >, or = in the oval.

14. 7 × 8 ◯ 9 × 6   15. 6 qt ◯ 2 gal

16. $7 ◯ 24 quarters

17. Mrs. Adams discovered that one pair of new glasses costs $212. Three of her children need glasses. Estimate how much money is needed for glasses.

18. Mrs. Anderson made 53 quarts of applesauce. She plans to put it in pint jars. How many jars does she need?

19. Twelve rows of soldiers marched by. There were four soldiers in each row. How many soldiers marched by?

20. A bucket holds 7 quarts, and a pitcher holds 3 quarts. Erica emptied the bucket twice and the pitcher three times. Cammi emptied the bucket once and the pitcher six times. Who poured more water? Write the amounts as an inequality (using < or >).

SYSTEMATIC REVIEW

22E

Round to the nearest 100.

1. 354 _____   2. 238 _____

3. 618 _____

Round the top number to the nearest 10 and estimate the answer.

4.   15 →
  × 4
  ―――

5.   69 →
  × 5
  ―――

6.   92 →
  × 6
  ―――

Round the top number to the nearest 100 and estimate the answer.

7.   182 →
  × 8
  ―――

8.   519 →
  × 4
  ―――

9.   393 →
  × 3
  ―――

SYSTEMATIC REVIEW 22E

Multiply using standard notation and place-value notation.

10. 4 4   4 0 + 4   11. 3 2   3 0 + 2
    × 2   ×   2   × 2   ×   2

12. 3 0 3   3 0 0 + 0 0 + 3   13. 1 2 2   1 0 0 + 2 0 + 2
    ×   3   ×   3   ×   4   ×   4

Write <, >, or = in the oval.

14. 8 × 8 ◯ 6 × 10   15. 40 cents ◯ 8 nickels

16. 5 gal ◯ 45 pts

17. Katherine picked 12 quarts of strawberries. How many pints of berries did she pick?

18. The next day, Katherine picked 23 more quarts of berries. Using your answer to #17, find how many pints of berries have been picked in all.

19. Christopher wants to make a border to go all around his room. His room is 13 ft long and 12 ft wide. He has made 50 ft of border. Does he have enough yet?

20. If Christopher (#19) had purchased the border instead of making it, it would have cost $1.00 per foot. He spent $19.00 on materials. How much did he save?

SYSTEMATIC REVIEW

22F

Round to the nearest 1,000.

1. 2,956 _____
2. 4,440 _____

3. 7,513 _____

Round the top number to the nearest 10 and estimate the answer.

4. 82 →
   × 7

5. 58 →
   × 3

6. 47 →
   × 4

Round the top number to the nearest 100 and estimate the answer.

7. 876 →
   ×   3

8. 691 →
   ×   2

9. 335 →
   ×   5

GAMMA SYSTEMATIC REVIEW 22F

SYSTEMATIC REVIEW 22F

Multiply using standard notation and place-value notation.

10. 13   10 + 3   11. 41   40 + 1
    × 3   ×   3   × 2   ×   2

12. 111   100 + 10 + 1   13. 323   300 + 20 + 3
    ×   8 ×   8   ×   3   ×   3

Write <, >, or = in the oval.

14. 3 × 8 ◯ 4 × 6   15. 56 cents ◯ 6 dimes

16. 3 × 12 ◯ 8 × 4

17. Jonathan was trapping for fur in the wilds of Wenham. Last winter, he caught 24 squirrels, 4 mountain lions, 3 coyotes, and 1 bison. How many animals did he catch?

18. The price for the skins (#17) was $2 per squirrel, $12 dollars per lion, $6 dollars per coyote, and $28 dollars per bison. How much money did Jonathan make?

19. A trucker uses 68 gallons of gasoline for one trip. Estimate how many gallons are needed to make six trips.

20. If the trucker in #19 hauls a lighter load, he needs only 54 gallons for the same trip. However, he will have to make the trip 7 times to get everything delivered. Estimate the number of gallons needed for 7 trips. Use an inequality to compare your answers to #19 and #20.

APPLICATION AND ENRICHMENT

22G

Color the picture. Complete each step in the order given for best results.
If you have already colored a number, do not color it again in the next step.

If you say the number when you skip count by 4 to 40, color the space red.
If you say the number when you skip count by 6 to 60, color the space green.
If you say the number when you skip count by 7 to 70, color the space blue.
If you say the number when you skip count by 8 to 80, color the space orange.

GAMMA APPLICATION AND ENRICHMENT 22G

APPLICATION AND ENRICHMENT 22G

You can use drawings and charts to answer questions about patterns.

The first week, the carpenter built a single square room for the new house. Each week after that he built rooms of the same size all around the walls of the house.

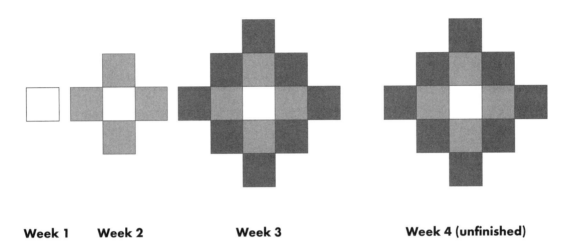

Week 1   **Week 2**   **Week 3**   **Week 4 (unfinished)**

1. How many rooms did the carpenter build during week 1? ___

2. How many rooms did the carpenter build during week 2? ___

3. How many rooms did the carpenter build during week 3? ___

Now fill in the chart to show what was built the first three weeks.

Weeks	1	2	3	4	5	6	7
Number of Rooms built	1						

4. Finish the drawing for week 4. Count the new rooms. How many rooms did the carpenter build during week 4? _____

5. Write your answer to #4 in the chart under week 4.

6. If you see the pattern, use it to finish filling in the chart without drawing any more rooms.

LESSON PRACTICE 23A

Build each rectangle and write the factors. Then group blocks with the same place value and write the product. The first one has been done for you.

1.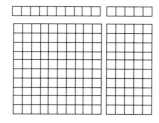

   The factors are __15__ × __11__ . The product is __165__ .

2.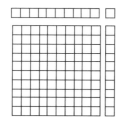

   The factors are ___ × ___ . The product is _____ .

Multiply using standard notation and place-value notation. These may be built with one set of blocks. The first one has been done for you.

3.
```
    14   →      10 + 4
   × 12  ↑    × 10 + 2
   ─────       ────────
     28         20 + 8
   + 14      + 100 + 40
   ─────      ─────────
    168       100 + 60 + 8
```

4.
```
    32   →      30 + 2
   × 11  ↑    × 10 + 1
   ─────       ────────
```

LESSON PRACTICE 23A

5. $\begin{array}{r} 22 \\ \times 10 \end{array} \rightarrow \begin{array}{r} 20+2 \\ \times 10+0 \end{array}$

6. $\begin{array}{r} 23 \\ \times 13 \end{array} \rightarrow \begin{array}{r} 20+3 \\ \times 10+3 \end{array}$

7. $\begin{array}{r} 12 \\ \times 12 \end{array} \rightarrow \begin{array}{r} 10+2 \\ \times 10+2 \end{array}$

8. $\begin{array}{r} 21 \\ \times 14 \end{array} \rightarrow \begin{array}{r} 20+1 \\ \times 10+4 \end{array}$

9. George made $12 profit on every chair he sold. If he sold 13 chairs, how much did he make?

10. Mrs. Softheart had 17 cats. If each cat had 11 kittens, how many new pets did she have?

LESSON PRACTICE 23B

Build each rectangle and write the factors. Then group blocks with the same place value and write the product.

1.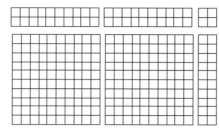

   The factors are ___ × ___ . The product is _____ .

2.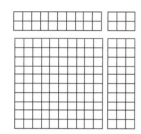

   The factors are ___ × ___ . The product is _____ .

Multiply using standard notation and place-value notation. These may be built with one set of blocks.

3.   13   →   10 + 3
       × 11   ↑   × 10 + 1

4.   21   →   20 + 1
       × 12   ↑   × 10 + 2

LESSON PRACTICE 23B

5. $\begin{array}{r} 45 \\ \times 11 \end{array} \rightarrow \uparrow \begin{array}{r} 40+5 \\ \times 10+1 \end{array}$

6. $\begin{array}{r} 23 \\ \times 21 \end{array} \rightarrow \uparrow \begin{array}{r} 20+3 \\ \times 20+1 \end{array}$

7. $\begin{array}{r} 22 \\ \times 13 \end{array} \rightarrow \uparrow \begin{array}{r} 20+2 \\ \times 10+3 \end{array}$

8. $\begin{array}{r} 37 \\ \times 10 \end{array} \rightarrow \uparrow \begin{array}{r} 30+7 \\ \times 10+0 \end{array}$

9. How many months are in 20 years? (If you do not already know that there are 12 months in a year, this is a good time to memorize that fact.)

10. My garden has 11 rows of corn with 41 plants in each row. How many corn plants are in my garden?

LESSON PRACTICE

23C

Build each rectangle and write the factors. Then group blocks with the same place value and write the product.

1.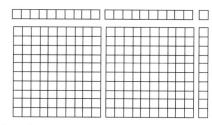

   The factors are ___ × ___ . The product is _____ .

2.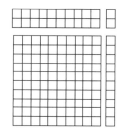

   The factors are ___ × ___ . The product is _____ .

Multiply using standard notation and place-value notation. These may be built with one set of blocks.

3. $22 \rightarrow 20+2$
   $\underline{\times 11}\uparrow\underline{\times 10+1}$

4. $33 \rightarrow 30+3$
   $\underline{\times 12}\uparrow\underline{\times 10+2}$

LESSON PRACTICE 23C

5. $\begin{array}{r} 11 \\ \times\,16 \end{array}$ → $\begin{array}{r} 10+1 \\ \times\,10+6 \end{array}$ ↑

6. $\begin{array}{r} 19 \\ \times\,10 \end{array}$ → $\begin{array}{r} 10+9 \\ \times\,10+0 \end{array}$ ↑

7. $\begin{array}{r} 22 \\ \times\,22 \end{array}$ → $\begin{array}{r} 20+2 \\ \times\,20+2 \end{array}$ ↑

8. $\begin{array}{r} 44 \\ \times\,11 \end{array}$ → $\begin{array}{r} 40+4 \\ \times\,10+1 \end{array}$ ↑

9. Ashlea spotted 15 chickadees sitting on each of 11 branches. How many birds did she spot in all?

10. What is the area of a room that is 13 feet long and 11 feet wide?

SYSTEMATIC REVIEW

Multiply.

1. 20
   × 13

2. 27
   × 11

3. 13
   × 13

4. 24
   × 21

5. 12
   × 31

6. 23
   × 12

7. 12
   × 44

8. 31
   × 21

Round to the nearest 100.

9. 619 _____

10. 283 _____

11. 752 _____

SYSTEMATIC REVIEW 23D

Round the top number to the nearest 10 and estimate the answer.

12. 15 →
    × 9

13. 34 →
    × 7

14. 28 →
    × 6

Subtract.

15. 519
    − 124

16. 863
    − 293

17. 611
    − 299

18. The ants went marching two by two. How many ants were in 113 rows?

19. Two hundred twelve salmon jumped over the dam each day on their way upstream. How many salmon went over the dam in four days?

20. How many pints are in 100 quarts of juice?

SYSTEMATIC REVIEW

Multiply.

1. 30
   × 13

2. 21
   × 20

3. 31
   × 12

4. 44
   × 22

5. 22
   × 33

6. 13
   × 11

7. 21
   × 22

8. 10
   × 21

Round to the nearest 10.

9. 91 ___

10. 39 ___

11. 27 ___

SYSTEMATIC REVIEW 23E

Round the top number to the nearest 100 and estimate the answer.

12. 651 →
     × 3

13. 125 →
     × 5

14. 290 →
     × 8

Subtract.

15. 116
     − 95

16. 700
     − 602

17. 558
     − 349

18. Amos had 44 tons of silage in one silo and 26 tons in his other silo. In one month, he fed 21 tons of silage to his cows. How many tons of silage does he have left?

19. Estimation may also be used for addition. Keith bought a coat for $39 and a pair of boots for $54. Round each number to the nearest 10 and estimate how much Keith spent in all.

20. Austin earns $60 a day. How much will he earn in 9 days?

SYSTEMATIC REVIEW

Multiply.

1. 35
   $\times 11$

2. 23
   $\times 10$

3. 26
   $\times 11$

4. 22
   $\times 13$

5. 44
   $\times 11$

6. 23
   $\times 12$

7. 12
   $\times 13$

8. 20
   $\times 44$

Round to the nearest 1,000.

9. 6,843 _____

10. 1,065 _____

11. 4,555 _____

SYSTEMATIC REVIEW 23F

Round the top number to the nearest 100 and estimate the answer.

12. 905 →
    ×   8

13. 613 →
    ×   9

14. 768 →
    ×   7

Add.

15.   298
    + 163

16.   562
    + 475

17.   809
    + 917

18. Peter Piper's pepper farm had 236 hot habanero pepper plants and 349 jalepeño pepper plants. After holding a pepper sale, he had 490 hot pepper plants left. How many plants had Peter Piper sold?

19. Margaret bought a case of pencils to give out at her booth at the fair. The pencils were packaged one dozen in a pack. There were 24 packs in the case. How many pencils did she have to give away? (Memorize: 1 dozen = 12)

20. If 230 people visited the zoo each day, how many people visited the zoo in three days?

APPLICATION AND ENRICHMENT

23G

Pretend that your favorite number is 11. Fill in the blanks and find the answers. Write each answer in the box under the problem.

1. You spent $11 every day to buy _____ .

   How much did you spend in 23 days?

   ☐

2. You made 11 _____ every hour.

   How many of them did you make in 17 hours?

   ☐

3. You ate 11 _____ at each meal.

   How many did you eat at 34 meals?

   ☐

4. You saw 11 more _____ every time you looked.

   If you looked 63 times, how many did you see altogether?

   ☐

APPLICATION AND ENRICHMENT 23G

Answer the questions. Connect the answers in order to find the picture.

1. How many pints make one quart?

2. How many feet make one yard?

3. How many quarts make one gallon?

4. How many cents make one nickel?

5. How many days make one week?

6. How many legs does a spider have?

7. How many pints make 12 quarts?

8. How many feet make 13 yards?

9. How many quarts make 11 gallons?

10. How many days make 10 weeks?

11. How many pints make 44 quarts?

12. How many feet make 33 yards?

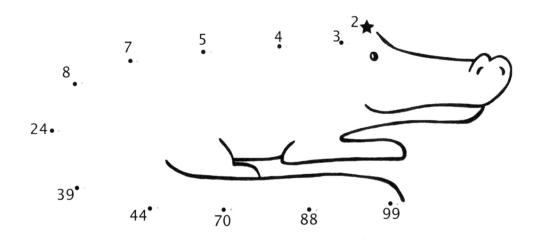

LESSON PRACTICE

Regroup and multiply. Check your work with place-value notation. The first one has been done for you.

1.
```
    14         10 + 4
   ×17        ×10 + 7
   ──────    ────────
   ②            ⑳
    78          70 + 8
   ①          ⑩⑩
   14         100 + 40
   ────       ────────────
   238        200 + 30 + 8
```

2.
```
    24
   ×18
   ────
```

3.
```
    22
   ×26
   ────
```

4.
```
    46
   ×12
   ────
```

5.
```
    27
   ×16
   ────
```

6.
```
    36
   ×24
   ────
```

LESSON PRACTICE 24A

7. 35
    $\underline{\times 29}$

8. 25
    $\underline{\times 23}$

9. The parking lot was full. If there were 15 rows with 35 cars in each, how many cars were in the parking lot?

10. The mill produced 25 tons of steel each day. How much did it produce in 12 days?

LESSON PRACTICE 24B

Regroup and multiply. Check your work with place-value notation.

1. 19
    × 32

2. 42
    × 66

3. 33
    × 26

4. 37
    × 12

5. 13
    × 19

6. 16
    × 29

LESSON PRACTICE 24B

7. 34   8. 48
   ×17   ×26

9. Andy earned $35 each week for 13 weeks. How much did he earn in all?

10. A barrel holds 45 gallons of water. If each gallon of water can support 24 mosquitoes, what is the possible number of mosquitoes that can live in the barrel?

LESSON PRACTICE 24C

Regroup and multiply. Check your work with place-value notation.

1. 23
   × 14

2. 27
   × 16

3. 29
   × 22

4. 35
   × 15

5. 28
   × 22

6. 36
   × 24

LESSON PRACTICE 24C

7. 44
    × 27

8. 56
    × 16

9. The dromedaries could each carry 75 pounds. If there were 13 dromedaries in the caravan, how many pounds could be transported?

10. Each person attending the convention paid a fee of $25. If 63 people attended, how much money was paid?

SYSTEMATIC REVIEW 24D

Multiply, regrouping if necessary.

1. 32
   × 17

2. 14
   × 28

3. 36
   × 22

4. 41
   × 38

5. 50
   × 32

6. 31
   × 33

7. 12
   × 41

8. 43
   × 12

Round the top number to the nearest 10 and estimate the answer.

9. 98 →
   × 4

10. 56 →
    × 8

11. 31 →
    × 7

SYSTEMATIC REVIEW 24D

Skip count and write the numbers.

12. ____, 4, ____, ____, 10, ____, ____, ____, ____, ____

Solve for the unknown and fill in the blanks.

13. 18 tsp = ___ Tbsp

14. 24 ft = ___ yd

15. 32 pt = ___ gal

16. A zoo has a pen for baby animals. If the pen was 15 feet long and 15 feet wide, how many square feet did the baby animals have in which to play?

17. If Derrick spends 13 hours a week at his computer, how much time does he spend in a year (52 weeks)?

18. Traveling down the river in her canoe, Elspeth drifted lazily along for 240 yards. When the river got rocky, she picked her way carefully for 335 yards. Then she entered the rapids and paddled furiously for another 124 yards. How many yards down the river did she travel in all?

19. A crossword puzzle had 61 clues down and 57 clues across. Steve figured out 59 of the clues. How many are left?

20. Karena filled two five-gallon buckets with water for her tomato plants. If she used one pint of water for each plant, how many plants could she water?

SYSTEMATIC REVIEW

24E

Multiply, regrouping if necessary.

1. 18
   × 40

2. 39
   × 23

3. 36
   × 99

4. 76
   × 89

5. 24
   × 21

6. 21
   × 14

7. 23
   × 22

8. 12
   × 13

Round the top number to the nearest 100 and estimate the answer.

9. 608 →
   × 3

10. 145 →
    × 5

11. 762 →
    × 8

Skip count and write the numbers.

12. ___, ___, 9, ___, ___, ___, ___, ___, ___, 30

Solve for the unknown and fill in the blanks.

13. 36 qt = ___ gal

14. 25 cents = ___ nickels

15. 12 pt = ___ qt

16. A king claimed a territory 47 miles long and 31 miles wide. How many square miles did he claim?

   The king gave 100 square miles of land to his son. How many square miles did the king have left?

17. Mom bought three dozen eggs. We already had two dozen eggs at home. How many eggs do we have now?

18. If Shirley bought 11 candy bars for 25 cents apiece, how many cents did she spend?

19. Sean swam 278 yards. Estimate to the nearest 100 to find about how many feet he swam.

20. When he made apple pies, Leo used 7 apples per pie. He cut each apple into 8 slices. How many slices of apple are needed for 3 pies? Multiply two ways to find the answer. Do the multiplication inside the parentheses first each time.

   $(7 \times 8) \times 3 =$ _____ $7 \times (8 \times 3) =$ _____

SYSTEMATIC REVIEW

24F

Multiply, regrouping if necessary.

1. 41
   × 62

2. 55
   × 25

3. 53
   × 35

4. 28
   × 38

5. 49
   × 38

6. 12
   × 42

7. 24
   × 22

8. 55
   × 11

Round the top number to the nearest 100 and estimate the answer.

9. 158 →
   × 4

10. 290 →
    × 9

11. 312 →
    × 3

Skip count and write the numbers.

12. ____, 8, ____, ____, 20, ____, ____, ____, ____, ____

Solve for the unknown and fill in the blanks.

13. 60 cents = ___ dimes

14. 28 quarters = $___

15. 15 ft = ___ yd

16. Naomi and her mom made two dozen cookies today. How many cookies did they make?

17. Silas likes trucks. His dad told him that large trucks have 18 wheels and small trucks have 4 wheels. How many wheels are on two large trucks and five small trucks?

18. Casey read 10 books a month for one year. How many books did he read in all that year?

19. During a basketball game, our team made 23 two-point shots and 3 three-point shots. How many points did we make in all?

20. The heating oil tank at Jill's house had 82 gallons of oil in it. A truck came and added 140 gallons. How many quarts of oil are in the tank now?

APPLICATION AND ENRICHMENT

24G

Ashley built a pattern with black and white squares. First she put down a white square, and then she put black squares all around it.

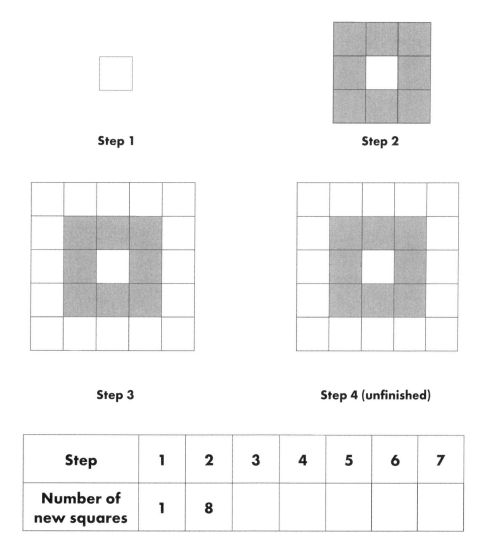

Step	1	2	3	4	5	6	7
Number of new squares	1	8					

1. Look at the picture to see how many Ashley put down for Step 3 and write the correct number in the chart.

2. Finish the drawing for Step 4. Count the new black squares and write the correct number in the chart.

3. What skip count pattern do you see?

4. Use the skip count pattern to finish filling in the chart.

APPLICATION AND ENRICHMENT 24G

Color the picture.

If the number in the space rounds to 20, color the space yellow.
If the number in the space rounds to 30, color the space purple.
If the number in the space rounds to 40, color the space blue.
If the number in the space rounds to 50, color the space green.
If the number in the space rounds to 60, color the space red.
If the number in the space rounds to 70, color the space orange.

LESSON PRACTICE 25A

Estimate your answer, regroup, and multiply. The first one has been done for you.

1.
```
    4 3 5  →    (4 0 0)
    × 2 5       × (3 0)
    1 2         (1 2,0 0 0)
    2 0 5 5
    1
    8 6 0
    1 0,8 7 5
```

2.
```
    6 2 4  →
    × 8 1
```

3.
```
    3 0 5  →
    × 2 1
```

4.
```
    3 1 9  →
    × 3 3
```

5.
```
    4 9 5  →
    × 7 2
```

6.
```
    8 7 6  →
    × 1 9
```

LESSON PRACTICE 25A

7. 352 →
   × 25

8. 681 →
   × 38

Estimate and solve each of these word problems.

9. Jeff bought 18 packages of toothpicks for a school project. Each box contained 500 toothpicks. How many toothpicks did Jeff buy?

   _____ , _____

10. Twenty-three cars finished the 642-mile cross-country race last year. How many miles had all the cars traveled altogether?

    _____ , _____

11. Mr. King has 212 milk cows in his herd. If they each produce 25 gallons of milk a day, how many gallons does the whole herd produce in a day?

    _____ , _____

12. One shelf of the library holds a set of books with 352 pages in each book. There are 15 books in the set. How many pages are on the shelf?

    _____ , _____

LESSON PRACTICE 25B

Estimate your answer, regroup, and multiply.

1. 125 →
   × 25

2. 681 →
   × 38

3. 492 →
   × 75

4. 145 →
   × 15

5. 534 →
   × 44

6. 719 →
   × 99

LESSON PRACTICE 25B

7. 154 →
    ×16

8. 290 →
    ×41

Estimate and solve each of these word problems.

9. Five hundred eighty-two mosquitoes laid their eggs. If each had 73 offspring who survived, how many new mosquitoes were there?

    _____ , _____

10. The city can buy lawn mowers for $235 apiece. Thirteen new mowers are needed to take care of the city parks. How much money should be budgeted for this expense?

    _____ , _____

11. Brandon put 35 pennies in a jar every day for a year (365 days). How many pennies did he have at the end of the year?

    _____ , _____

12. If Brandon puts in two times as many pennies each day (#11), how many pennies will he have at year's end?

    _____ , _____

LESSON PRACTICE 25C

Estimate your answer, regroup, and multiply.

1. 816 →
   × 79

2. 492 →
   × 55

3. 373 →
   × 64

4. 777 →
   × 33

5. 436 →
   × 36

6. 947 →
   × 14

LESSON PRACTICE 25C

7. 559 →
   × 63

8. 519 →
   × 52

Estimate and solve each of these word problems.

9. What is the area of a rectangular field that is 963 feet long and 98 feet wide?

   _____, _____

10. Margaret earns $279 a week. What does she earn in 52 weeks?

    _____, _____

11. A school ordered 105 dozen pencils. How many pencils did it receive when the order came?

    _____, _____

12. On his trip across the United States, Tim drove 215 miles a day for two weeks. How far did he drive in all?

    _____, _____

SYSTEMATIC REVIEW

25D

Regroup and multiply.

1. 873
   × 30

2. 718
   × 38

3. 314
   × 35

4. 85
   × 36

5. 42
   × 59

6. 61
   × 47

Find the area and perimeter of each rectangle.

Rectangle: 33 ft by 11 ft

7. A = _____

8. P = _____

Rectangle: 14 in by 12 in

9. A = _____

10. P = _____

Square: 25 in by 25 in

11. A = _____

12. P = _____

GAMMA SYSTEMATIC REVIEW 25D

SYSTEMATIC REVIEW 25D

Write <, >, or = in the oval.

13. 6 × 7 ◯ 5 × 8

14. 3 × 12 ◯ 9 × 4

15. 3 × 5 ◯ 8 + 8

Skip count and write the numbers.

16. ___, ___, 15, ___, 25, ___, ___, ___, ___, ___

17. Elizabeth is a fine dishwasher. Her dad figured she washes 36 pieces of silverware after each meal. If she washes the dishes after each meal this week (3 × 7), how many pieces of silverware will she have washed?

18. Estimate the cost of going to college for four years if each year costs $9,600. (Round to the nearest thousand first.)

19. The parking lot at Jayne's car dealership has room for 55 cars. She has 43 cars on the lot and is expecting 18 more to be delivered. How many cars does she need to sell to make room for the ones that are coming?

20. Matthew peered through the water as he swam in the lake, but he could see only 12 yards ahead. How many feet through the water could Matthew see?

SYSTEMATIC REVIEW

Regroup and multiply.

1. 522
    × 93

2. 832
    × 57

3. 471
    × 84

4. 347
    ×   8

5. 43
    × 25

6. 24
    × 31

Find the area and perimeter of each rectangle.

```
        40 ft
21 ft [        ] 21 ft
        40 ft
```

7. A = _____

8. P = _____

```
        13 in
7 in [      ] 7 in
        13 in
```

9. A = _____

10. P = _____

```
        15 in
15 in [    ] 15 in
        15 in
```

11. A = _____

12. P = _____

SYSTEMATIC REVIEW 25E

Write <, >, or = in the oval.

13. 9 × 9 ◯ 2 × 40   14. 6 × 8 ◯ 90 − 40

15. 11 × 6 ◯ 8 × 8

Skip count and write the numbers.

16. ___, ___, ___, ___, ___, 36, ___, 48, ___, ___

17. A fisherman catches about 962 fish a day. About how many fish does he catch in a week if he takes one day off to rest? Estimate and then solve.

    _____, _____

18. Brandi sold seven gallons of homemade root beer in one day. If the root beer was served in one-pint mugs, how many servings did she sell?

19. Michelle has 21 bags of grain for her animals in the back of her pickup truck. If each bag of grain weighs 100 pounds, what is the total weight of her load?

20. Four hundred fifty-five people bought tickets for the play. If the tickets cost $15 apiece, how much money was taken in?

    Thirty-five people stayed home because they were sick. How many people actually saw the play?

SYSTEMATIC REVIEW

Regroup and multiply.

1. 712
    × 65

2. 360
    × 58

3. 252
    × 38

4. 432
    ×   7

5.   41
    × 15

6.   62
    × 23

Find the area and perimeter of each rectangle.

```
        38 ft
16 ft [        ] 16 ft
        38 ft
```

7. A = _____

8. P = _____

```
        15 in
10 in [      ] 10 in
        15 in
```

9. A = _____

10. P = _____

```
       12 in
12 in [  ] 12 in
       12 in
```

11. A = _____

12. P = _____

SYSTEMATIC REVIEW 25F

Write <, >, or = in the oval.

13. 6 × 6 ◯ 4 × 9

14. 12 × 12 ◯ 100 + 50

15. 2 × 11 ◯ 11 + 11

Skip count and write the numbers.

16. ____, 14, ____, ____, ____, ____, ____, ____, 63, ____

17. Esther is 98 years old today. For how many months has she lived?

18. Deb took 11 dollars to the bank and exchanged them for quarters. How many more quarters will she need to make a total of 50 quarters?

19. Sara sent out 375 copies of her newsletter by e-mail and 166 copies by snail mail. How many copies has she sent out altogether? Estimate by rounding the numbers to the nearest hundred first and then find an exact answer.

   _____, _____

20. Nena couldn't wait to see Uncle Peter, who was coming in exactly 4 weeks. Since there are 7 days in a week and 24 hours in a day, how many hours did she have to wait? Multiply two ways to find the answer.

   (4 × 7) × 24 = _____

   4 × (7 × 24) = _____

APPLICATION AND ENRICHMENT

25G

Use the words and clues to fill in the crossword puzzle.

- add
- double
- hundreds
- multiplicand
- multiplier
- product
- tens
- ten thousands (enter as one word)
- thousands
- unit

Across

2. The bottom factor in a multiplication problem is the _____.

4. In 56,378, the 5 is in the _____ place.

5. The answer to a multiplication problem is the _____.

7. In 23,541, the 5 is in the _____ place.

8. When multiplying greater numbers, _____ the partial products.

9. _____-digit multiplication is more than one problem.

Down

1. In 65,982, the 5 is in the _____ place.

2. The top number in a multiplication problem is the _____.

3. The number 31 means three tens and one _____.

6. In 14,356, the 5 is in the _____ place.

APPLICATION AND ENRICHMENT 25G

Color the picture.

If the number in the space rounds to 200, color the space red.
If the number in the space rounds to 300, color the space yellow.
If the number in the space rounds to 400, color the space blue.
If the number in the space rounds to 500, color the space green.
If the number in the space rounds to 600, color the space orange.
If the number in the space rounds to 700, color the space purple.

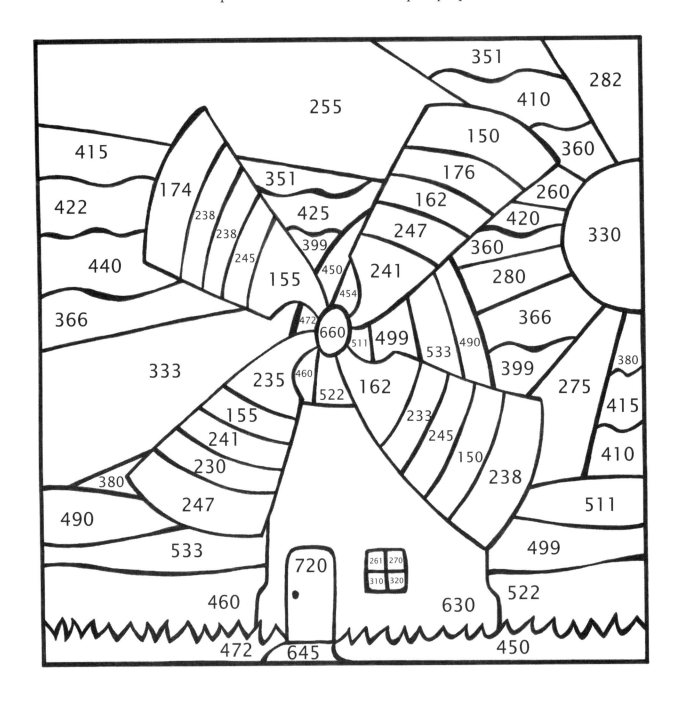

LESSON PRACTICE 26A

Build rectangles and find all the possible pairs of factors. The first one has been done for you.

1. 8 $\underline{1} \times \underline{8}$
       $\underline{2} \times \underline{4}$

2. 12   ___ × ___
        ___ × ___
        ___ × ___

3. 4   ___ × ___
       ___ × ___

4. 15   ___ × ___
        ___ × ___

5. 10   ___ × ___
        ___ × ___

6. 21   ___ × ___
        ___ × ___

7. 14   ___ × ___
        ___ × ___

8. 9   ___ × ___
       ___ × ___

9. 6   ___ × ___
       ___ × ___

LESSON PRACTICE 26A

10. How many pennies equal eight quarters?

11. Lauren found 12 quarters on the playground. How many cents did she find?

12. It took three quarters to get a drink from the vending machine. What was the price of the drink?

13. Patty is planting 12 peonies in her garden. Describe all the ways she can arrange them to form a rectangle.

    ____ × ____ , ____ × ____ , ____ × ____

14. Joshua has 16 pennies. List the rectangular arrangements he can make with them.

    ____ × ____ , ____ × ____ , ____ × ____

15. Alisha bought 18 decorative tiles for her kitchen wall. What are the different ways she can arrange them to form a rectangle?

    ____ × ____ , ____ × ____ , ____ × ____

LESSON PRACTICE 26B

Build rectangles and find all the possible pairs of factors.

1. 16 ___ × ___
       ___ × ___
       ___ × ___

2. 18 ___ × ___
       ___ × ___
       ___ × ___

3. 22 ___ × ___
       ___ × ___

4. 6   ___ × ___
       ___ × ___

5. 14 ___ × ___
       ___ × ___

6. 15 ___ × ___
       ___ × ___

7. 8   ___ × ___
       ___ × ___

8. 12 ___ × ___
       ___ × ___
       ___ × ___

9. 10 ___ × ___
       ___ × ___

LESSON PRACTICE 26B

10. Tom has nine quarters in his piggy bank. He would like to exchange them to get pennies. How many pennies should Tom receive?

11. Amy found 16 quarters she didn't know she had. How many extra cents does she have?

12. Mom found 35 quarters in the washing machine over the last month. How many cents did she find?

13. Justin has 20 dominoes. Describe the different rectangles he can make.

    ____ × ____ , ____ × ____ , ____ × ____

14. George wants to plant 21 apple trees. If each row must be the same length, list the ways he can arrange his orchard.

    ____ × ____ , ____ × ____

15. Pretend that the United States has only 24 states. What are the possible ways the stars can be arranged on the flag if all the rows must have the same number of stars?

    ____ × ____ , ____ × ____ , ____ × ____ , ____ × ____

LESSON PRACTICE 26C

Build rectangles and find all the possible pairs of factors.

1. 22   ___ × ___
       ___ × ___

2. 8   ___ × ___
       ___ × ___

3. 6   ___ × ___
       ___ × ___

4. 14   ___ × ___
       ___ × ___

5. 21   ___ × ___
       ___ × ___

6. 10   ___ × ___
       ___ × ___

7. 18   ___ × ___
       ___ × ___
       ___ × ___

8. 9   ___ × ___
       ___ × ___

9. 4   ___ × ___
       ___ × ___

LESSON PRACTICE 26C

10. Alex plans to trade his pennies for quarters. How many pennies will he need to get four quarters?

11. Drew's dad gave him a quarter every day for two weeks. How many cents did Drew get?

12. Molly has five quarters. Does she have enough money to buy a toy that costs 95¢?

13. The drill team has 16 members. Describe the rectangular arrangements they can make.

14. If the team (#13) gains four new members, how many rectangles will they be able to make?

    Describe all the arrangements they can make now.

15. Zarah is playing with Math-U-See blocks. List the different rectangles she can make that have an area of 12 blocks.

SYSTEMATIC REVIEW

26D

Find all the possible pairs of factors.

1. 15 ____ × ____

   ____ × ____

2. 9 ____ × ____

   ____ × ____

3. 4 ____ × ____

   ____ × ____

Multiply.

4. 11 quarters = ____ cents

5. 6 quarters = ____ cents

6. 21 quarters = ____ cents

Regroup and multiply.

7. 4 2 3
    × 5 7

8. 2 7 6
    × 1 2

9. 6 1 4
    × 3 2

10. 1 3 4
    ×   4

11.   7 4
    × 3 3

12.   5 1
    × 1 6

SYSTEMATIC REVIEW 26D

Subtract.

13.   3 9 7
    −   6 3

14.   4 1 1
    − 3 5 0

15.   6 0 0
    − 1 0 4

Fill in the blanks in the numerators and denominators to name the equivalent fractions.

16. $\dfrac{1}{2} = \dfrac{}{} = \dfrac{3}{} = \dfrac{}{} = \dfrac{}{10}$

17. A garden catalog says that a certain set of plants will fill 20 square feet. What are the different sizes Mary could make her rectangular garden, using whole feet on a side?

18. A store manager bought 35 chairs for $22 apiece. How much did he spend on chairs?

19. Alex counted 231 cars on his way to town. If his brother counted the same number going the other way, how many did they count in all?

20. Tara had 13 quarters. Her mother gave her six more quarters. How many cents does she have?

SYSTEMATIC REVIEW

26E

Find all the possible pairs of factors.

1. 6   ___ × ___
         ___ × ___

2. 12   ___ × ___
         ___ × ___
         ___ × ___

3. 21   ___ × ___
         ___ × ___

Multiply.

4. 7 quarters = ___ cents

5. 10 quarters = ___ cents

6. 15 quarters = ___ cents

Regroup and multiply.

7. 125
   × 54

8. 731
   × 18

9. 378
   × 49

10. 276
   ×   ?

11. 38
   × 12

12. 25
   × 39

SYSTEMATIC REVIEW 26E

Subtract.

13. 554
    − 27

14. 976
    − 763

15. 483
    − 299

Fill in the blanks in the numerators and denominators to name the equivalent fractions.

16. $\dfrac{2}{3} = \dfrac{\_}{6} = \dfrac{\_}{\_} = \dfrac{8}{\_} = \dfrac{\_}{\_}$

17. There were 321 sets of triplets at the triplet convention. How many people were at the convention?

18. Taylor ran for 300 yards. How many feet did he run?

19. Sam has 15 green unit pieces. How many different rectangular arrangements can he make with them? List the pairs of factors shown by the rectangles.

20. Rhonda sneezed quietly 11 times and then sneezed loudly 3 more times. Her brother sneezed only 5 times. What is the difference between the number of times Rhonda sneezed and the number of times her brother sneezed?

SYSTEMATIC REVIEW

Find all the possible pairs of factors.

1. 24   ___ × ___
         ___ × ___
         ___ × ___
         ___ × ___

2. 16   ___ × ___
         ___ × ___
         ___ × ___

3. 10   ___ × ___
         ___ × ___

Multiply.

4. 2 quarters = ___ cents

5. 17 quarters = ___ cents

6. 20 quarters = ___ cents

Regroup and multiply.

7. 249
   × 12

8. 218
   × 75

9. 862
   × 10

10. 172
    × 6

11. 92
    ×17

12. 56
    ×24

SYSTEMATIC REVIEW 26F

Subtract.

13. 276
    − 12

14. 554
    − 396

15. 672
    − 325

Fill in the blanks in the numerators and denominators to name the equivalent fractions.

16. $\frac{5}{6} = \frac{}{} = \frac{}{18} = \frac{}{} = \frac{25}{}$

17. Ruth just finished making 24 gallons of apple juice, which she wants to put into quart jars. How many jars does she need?

18. How many quarters are needed to buy a book that costs $12?

19. Our backyard measures 125 feet by 45 feet. What is the area of the yard?

20. Our house sits along one short side of the backyard. (#19) How many feet of fence do we need to go around the other three sides of the yard?

APPLICATION AND ENRICHMENT

26G

Help to write the word problems by choosing numbers and nouns. (A noun is the name of a person, place, or thing.) One number in each problem should have three digits. The other number can have one or two digits.

1. Ryan picked _____ _____ from his garden
   (number)   (noun)
   every day for _____ days. How many of them did he
                 (number)
   pick altogether? _____
                    (answer)

2. Rose made _____ _____ every week
              (number)   (noun)
   for _____ weeks. How many did she make
       (number)
   altogether? _____
               (answer)

3. Lori cooked _____ _____ every
                (number)   (noun)
   day for _____ days. How many did Lori cook
           (number)
   altogether? _____
               (answer)

4. In one hour, Ethan found _____ new _____.
                             (number)   (noun)
   How many did he find in _____ hours? _____
                           (number)   (answer)

5. I saw _____ crawl up the wall in groups
         (noun)
   of _____ If there were _____ groups, how many
      (number)   (number)
   crawled up the wall? _____
                        (answer)

APPLICATION AND ENRICHMENT 26G

You have learned how to find the factors of a number by building rectangles. There are some simple rules that can help you find factors without building.

1. Two is a factor if the number is even. Even numbers are the ones we say when we skip count by two. Circle the numbers that have two for a factor.

       4   5   8   17   22

2. Five is a factor if the number has a five or zero in the units place. Circle the numbers that have five for a factor.

       10   25   31   40   56

3. Ten is a factor if the number has a zero in the units place. Circle the numbers that have 10 for a factor.

       30   16   45   50   72

4. Three is a factor if the digits in that number add up to three or a number that we say when we skip count by three. For example, we know that three is a factor of 24 because 2 + 4 = 6, and we say six when we skip count by three. Circle the numbers that have three for a factor.

       28   33   51   69   85

5. Nine is a factor if the digits in a number add up to nine or a number that we say when we skip count by nine. Circle the numbers that have nine for a factor.

       36   64   99   108   234

LESSON PRACTICE 27A

Say each number and then write it out in words. The first one has been done for you.

1. 2,345,172

   Two million, three hundred forty-five thousand, one hundred seventy-two

2. 261,829,130

   _____

   _____

Write each number in standard notation. The first one has been done for you.

3. 3,000 + 200 + 50 + 9 = 3,259

4. 40,000 + 2,000 + 300 + 10 + 6 _____

5. 100,000 + 40,000 + 9,000 + 200 + 70 + 3 _____

6. 2,000,000 + 100,000 + 30,000 + 4,000 + 900 + 10 + 1 _____

GAMMA LESSON PRACTICE 27A

LESSON PRACTICE 27A

Write each number in place-value notation. The first one has been done for you.

7. 72,375,100

   70,000,000 + 2,000,000 + 300,000 + 70,000 + 5,000 + 100

8. 150,941,220 _____

9. 600,400,090 _____

Multiply to find the number of ounces.

10. 6 pounds = ___ ounces   11. 10 lb = ___ oz

12. 13 lb = ___ oz

13. Mackenzie weighs 100 pounds. How many ounces does Mackenzie weigh?

14. A baby weighs eight pounds. How many ounces does that baby weigh?

15. Which weighs more, a two-pound can of spaghetti or a 30-ounce can of spaghetti?

LESSON PRACTICE 27B

Say each number and write it out in words.

1. 16,704,900 _____

   _____

2. 321,954,000 _____

   _____

Write each number in standard notation.

3. 4,000 + 300 + 80 _____

4. 300,000 + 40,000 + 9,000 + 600 + 20 + 2 _____

5. 2,000,000 + 400,000 + 60,000 + 1,000 + 800 _____

6. 900,000,000 + 1,000 + 300 + 70 + 3 _____

LESSON PRACTICE 27B

Write each number in place-value notation.

7. 11,691,000 _____

8. 509,432,005 _____

9. 451,698,123 _____

Multiply to find the number of ounces.

10. 9 pounds = ___ ounces

11. 11 lb = ___ oz

12. 34 lb = ___ oz

13. Jason bought a five-pound box of chocolates for his wife. How many ounces of candy did she get?

14. Seth was excited because he had caught a seven-pound fish. How many ounces did his fish weigh?

15. We had a 23-pound turkey for Thanksgiving dinner. How many ounces did the turkey weigh?

LESSON PRACTICE

27C

Say each number and write it out in words.

1. 318,611,353 _____

   _____

2. 126,932 _____

   _____

Write each number in standard notation.

3. 20,000 + 3,000 + 900 + 10 + 4 _____

4. 70,000,000 + 5,000,000 + 100,000 + 50,000 + 4,000 + 900

   _____

5. 6,000,000 + 300 + 40 + 2 _____

6. 900,000,000 + 10,000,000 + 5,000,000 + 400,000 + 10,000 + 2,000 + 900 + 60 + 5

   _____

LESSON PRACTICE 27C

Write each number in place-value notation.

7. 321,618,000 _____

8. 30,500,800 _____

9. 101,007,003 _____

Multiply to find the number of ounces.

10. 4 pounds = ___ ounces

11. 12 lb = ___ oz

12. 51 lb = ___ oz

13. How many ounces are in a 10-pound bag of flour?

14. Mom bought 14 pounds of chicken on sale. How many ounces of chicken did she get?

15. A man weighs 212 pounds. How many ounces does he weigh?

SYSTEMATIC REVIEW 27D

Follow the directions.

1. Write in words: 10,650,300 _____

   _____

2. Write in standard notation: 600,000,000 + 30,000,000 + 2,000,000 + 100,000 + 70,000 + 8,000 + 400 + 30 + 1

   _____

3. Write in place-value notation: 456,789,000

   _____

Find all the possible pairs of factors.

4. 14   ____ × ____

       ____ × ____

5. 18   ____ × ____

       ____ × ____

       ____ × ____

6. 24   ____ × ____

       ____ × ____

       ____ × ____

       ____ × ____

Multiply and fill in the blanks.

7. 15 lb ____ oz

8. 19 quarters = ____ cents

9. 9 Tbsp = ____ tsp

SYSTEMATIC REVIEW 27D

Regroup and multiply.

10. 123
    × 67

11. 147
    × 51

12. 38
    × 15

Find the missing multiples of 2 and 8 in the equivalent fractions.

13. $\dfrac{2}{8} = \dfrac{}{} = \dfrac{}{24} = \dfrac{}{} = \dfrac{}{} = \dfrac{}{} = \dfrac{14}{} = \dfrac{}{} = \dfrac{}{} = \dfrac{}{}$

14. Nathaniel likes baseball. If there were seven teams in each division and nine starting players on each team, how many starters would be in one division?

15. Jill took a jar of quarters to the bank to be exchanged for dollar bills. If she received 16 dollars back, how many quarters were in the jar?

16. Lindsay had 15 Easter eggs to color. She colored 6 eggs red and 8 eggs blue. How many eggs were left to color?

17. Rob rode his bike 29 miles the first day of his trip. If he rode the same number of miles each day, estimate how far he could ride in 12 days.

18. Valerie went to the butcher shop and bought 5 pounds and 6 ounces of ground beef. How many ounces of ground beef did she buy?

SYSTEMATIC REVIEW

27E

Follow the directions.

1. Write in place-value notation: 356,000,000

   _____

2. Write in standard notation: 700,000,000 + 80,000,000 + 4,000,000 + 900,000

   _____

3. Write in words: 400,000,098

   _____

Find all the possible pairs of factors.

4. 16   ____ × ____   5. 10   ____ × ____

     ____ × ____   ____ × ____

     ____ × ____

6. 6   ____ × ____

     ____ × ____

Multiply and fill in the blanks.

7. 3 lb = ____ oz   8. $20 = ____ quarters

9. 13 yd = ____ ft

SYSTEMATIC REVIEW 27E

Regroup and multiply.

10. 557
    ×   3

11. 137
    × 59

12. 873
    × 21

Find the missing multiples of 3 and 9 in the equivalent fractions.

13. $\dfrac{3}{9} = \dfrac{}{} = \dfrac{}{} = \dfrac{}{36} = \dfrac{}{} = \dfrac{18}{} = \dfrac{}{} = \dfrac{}{} = \dfrac{}{} = \dfrac{}{}$

14. A carton holds 36 cans of soup. If the grocer has 43 cartons of soup, how many cans does he have altogether?

15. Austin gained 20 pounds this year. Then he got sick and lost 12 pounds. How many ounces lighter is he now?

16. Mary spends $175 a week on groceries. Estimate how much she spends in a month (four weeks).

17. There are 52 weeks in a year. How much would Mary spend on groceries in a year if she spent $175 every week?

18. The librarian checked out 95 books on Monday, 68 on Tuesday, 84 on Wednesday, 73 on Thursday, and 91 on Friday. How many books were checked out in all?

SYSTEMATIC REVIEW

Follow the directions.

1. Write in place-value notation: 400,400,400

   _____

2. Write in standard notation: 100,000,000 + 30,000,000 + 2,000,000 + 600,000 + 70,000 + 2,000 + 500 + 40 + 7

   _____

3. Write in words: 698,000,000

   _____

Find all the possible pairs of factors.

4. 9 ____ × ____   5. 15 ____ × ____

   ____ × ____   ____ × ____

6. 22 ____ × ____

   ____ × ____

Multiply and fill in the blanks.

7. 20 lb = ____ oz   8. 7 gal = ____ pints

9. 7 gal = ____ qt

SYSTEMATIC REVIEW 27F

Regroup and multiply.

10. 4 1 2
    × 2 4

11. 6 2 8
    × 4 1

12.   1 7
    × 5 2

Find the missing multiples of 4 and 10 in the equivalent fractions.

13. $\dfrac{4}{10} = \dfrac{}{} = \dfrac{}{} = \dfrac{}{} = \dfrac{}{60} = \dfrac{}{} = \dfrac{32}{} = \dfrac{}{} = \dfrac{}{}$

14. A baker's dozen is 13. If a baker sold 11 dozen rolls, how many rolls did he sell?

15. My square room measures 12 feet on each side. What is the area of the room? What is its perimeter?

16. Jack hoped his car had enough fuel to make it to the nearest gas station, which was 30 miles away. If the car ran out of fuel after 24 miles, how far did Jack have to walk?

17. Brenda bought five cases of ginger ale. Each case contained 24 12-ounce cans. How many ounces of ginger ale did Brenda buy? Multiply two ways to find the answer.

    (___ × ___) × ___ = _____   ___ × (___ × ___) = _____

18. Sam got these points for school projects: 95, 42, 68, and 81. How many points did he earn altogether?

APPLICATION AND ENRICHMENT

27G

Use the words and clues to fill in the crossword puzzle.

- dime
- dollar
- feet
- gallon
- nickel
- ounces
- tablespoon
- two

Across

3. There are four quarters in a _____ .

4. Sixteen _____ make one pound.

6. There are _____ pints in a quart.

7. There are three _____ in a yard.

Down

1. Four quarts make a _____ .

2. A _____ is the same amount as three teaspoons.

3. There are 10 cents in a _____ .

5. A coin worth five cents is a _____ .

GAMMA APPLICATION AND ENRICHMENT 27G

You can find the areas of larger shapes by adding up the parts. Look at the two rectangles in the example.

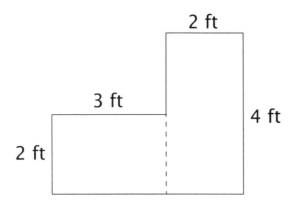

One rectangle measures 2 × 3, so its area is 6 square feet.
The other rectangle measures 2 × 4, so its area is 8 square feet.
Add the areas of the two rectangles to find the total area.
6 + 8 = 14 square feet

1. Find the area of the house.

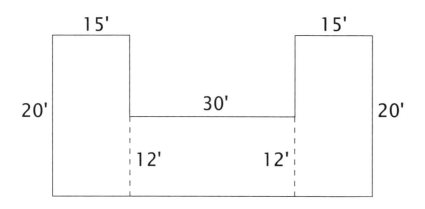

Area = _____ sq ft

If you wish, you may draw rooms or furniture in your house.

LESSON PRACTICE 28A

Multiply. Use estimation to see whether your answer makes sense. If you need more room for these problems, try turning a piece of notebook paper sideways and using the lines. The first one has been done for you.

1. 946 → (900)
    × 221   × (200)
           (180,000)

    946
    1882
    1882
    209,066

2. 284 →
    × 362

3. 880 →
    × 153

4. 714 →
    × 602

5. 1602 →
    ×   5

6. 1768 →
    ×   12

LESSON PRACTICE 28A

7.
```
| 8 | 1 | 7 | 2 | →
|   | × | 3 | 5 | 4
```

8.
```
| 4 | 6 | 7 | 5 | →
|   | × | 2 | 9 | 2
```

9. Marlene bought 120 copies of a new book to sell in her bookstore. If the book contains 325 pages, how many pages will Marlene have?

10. There are 1,440 minutes in a day. How many minutes total are in a week?

LESSON PRACTICE 28B

Multiply. Use estimation to see whether your answer makes sense.

1. 3 2 5 →
    × 2 1 3

2. 1 6 2 →
    × 5 4 8

3. 5 3 6 →
    × 1 3 4

4. 3 2 2 →
    × 7 2 5

5. 6 4 2 4 →
    ×   4

6. 3 4 4 5 →
    ×   9 3

LESSON PRACTICE 28B

7. 5|6|2|7 →
    ×|3|1|5

8. 3|5|7|9 →
    ×|4|6|2

9. If Dawn earned $2,450 a month, how much would she earn in a year?

10. The tires on Lisa's car go around 1,350 times for every mile she drives. How many times will each tire go around if she drives 396 miles?

LESSON PRACTICE 28C

Multiply. Use estimation to see whether your answer makes sense.

1. 627 × 450 →

2. 334 × 702 →

3. 234 × 121 →

4. 415 × 378 →

5. 3567 × 8 →

6. 2317 × 64 →

LESSON PRACTICE 28C

7. | 6 | 5 | 3 | 6 | →
    | × | 1 | 2 | 1 |

8. | 1 | 5 | 6 | 2 | →
    | × | 2 | 3 | 1 |

9. There are 3,600 seconds in an hour and 24 hours in a day. How many seconds are in a day?

10. Warren has a field that is 1,260 feet long and 845 feet wide. What is the area of his field?

SYSTEMATIC REVIEW

Multiply.

1. 125
    × 306

2. 7256
    ×   43

3. 8761
    × 280

Write in place-value notation.

4. 123,600,000 _____

Find all the possible pairs of factors.

5. 8   ___ × ___
       ___ × ___

6. 4   ___ × ___
       ___ × ___

7. 12 ___ × ___
       ___ × ___
       ___ × ___

8. 24 ___ × ___
       ___ × ___
       ___ × ___
       ___ × ___

SYSTEMATIC REVIEW 28D

Solve for the unknown.

9. $8D = 64$

10. $7X = 42$

11. $3F = 3$

12. $9B = 45$

13. A football field is 100 yards long and has a 10-yard end zone at each end. What is the total length in feet?

14. A pint of juice weighs one pound. What is the weight of three quarts of juice?

15. Sandi's recipe calls for five tablespoons of honey. All she has is a teaspoon for measuring. How many teaspoons of honey are needed?

16. Claire's backyard has a stream running through it. If 325 gallons of water flow through her backyard every minute, how many gallons flow through in an hour? (There are 60 minutes in an hour.)

17. Last year, Ellen sold 3,572 widgets at $213 apiece. How much money did she get for her widgets last year?

18. Driving through the countryside in Nova Scotia, Emily saw a band of 144 pipers playing their bagpipes. A while later, she saw 50 more pipers. Just as she got home, she was surprised to see 203 more pipers playing in her front yard. How many pipers did Emily see altogether?

SYSTEMATIC REVIEW

Multiply.

1. 433 × 127

2. 8192 × 74

3. 6123 × 245

Write in words.

4. 9,551,000 _____

Find all the possible pairs of factors.

5. 22   ___ × ___
      ___ × ___

6. 18   ___ × ___
      ___ × ___
      ___ × ___

7. 14   ___ × ___
      ___ × ___

8. 10   ___ × ___
      ___ × ___

SYSTEMATIC REVIEW 28E

Solve for the unknown.

9. 9D = 81

10. 6X = 54

11. 4F = 20

12. 8B = 0

13. Andrew built a wall with stones that weighed 125 pounds apiece. He used 1,200 stones in his wall. How many pounds did the wall weigh?

14. How many ounces does one of Andrew's stones weigh? (#13)

15. Thomas adds 43 stamps to his collection each month. How many stamps does he add in a year?

16. Hannah's bedroom measures 12 ft by 15 ft. How many one-foot-square tiles are needed to cover the floor?

17. The tiles Hannah likes cost $2 apiece. (#16) What will it cost to cover her floor?

18. William told 18 stories on Monday and 26 stories on Tuesday. Twenty-nine of William's stories were true. How many of his stories were not true?

SYSTEMATIC REVIEW

28F

Multiply.

1. | | 1 | 5 | 6 |
    |---|---|---|---|
    | × | | 5 | 2 | 3 |

2. | | 7 | 4 | 8 | 1 |
    |---|---|---|---|---|
    | × | | | 2 | 7 |

3. | | 1 | 2 | 2 | 2 |
    |---|---|---|---|---|
    | × | | | 4 | 4 | 3 |

Write in standard notation.

4. 60,000,000 + 5,000,000 + 900,000 + 10,000 _____

Find all the possible pairs of factors.

5. 9   ____ × ____

6. 20   ____ × ____

           ____ × ____

           ____ × ____

           ____ × ____

7. 15   ____ × ____

8. 21   ____ × ____

           ____ × ____

           ____ × ____

SYSTEMATIC REVIEW 28F

Solve for the unknown.

9. $7D = 56$

10. $8X = 72$

11. $5F = 50$

12. $6B = 36$

13. Ruth bought 13 bags of apples. Each bag contained 13 apples. How many apples did she buy?

14. There are 60 seconds in a minute and 60 minutes in an hour. How many seconds are in an hour?

15. Sherri bought 30 yards of ribbon to make bows for her party. Each bow uses one foot of ribbon. How many bows can she make in all?

16. A gallon of water weighs 128 ounces. How many ounces of water would a 275-gallon tank hold?

17. Mom bought one package of meat that weighed 5 pounds and one that weighed 2 pounds. She used 3 pounds for dinner. How many ounces does she have left?

18. Ed cleaned 116 windows in one building. There are 21 more buildings just like it. Round the number of windows to the nearest 100 and the number of buildings to the nearest 10. Estimate how many more windows he has to wash.

APPLICATION AND ENRICHMENT

This line graph shows you how a baby gained weight during her first year.

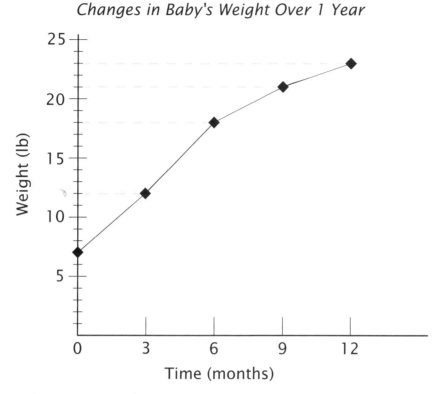

Use the graph to answer the questions.

1. How much did the baby weigh when she was born?

2. How much did the baby weigh when she was one year (12 months) old?

3. How much weight did the baby gain in one year?

4. Did the baby gain weight faster in the first half or the last half of the year?

5. How much did the baby weigh when she was six months old? Give your answer in ounces.

APPLICATION AND ENRICHMENT 28G

A bar graph can be used to compare the weights of different people in the same family.

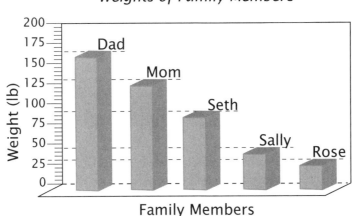

1. Study the scale at the left of the graph. How many pounds are represented by each small line?

2. Which person in the family weighs the least?

3. Which person weighs twice as much as Sally?

4. Mom weighed herself with Rose in her arms. What is their combined weight?

5. Do Mom and Rose together weigh as much as Dad?

6. Use an inequality sign to compare the combined weight of Mom and Seth with Dad's weight.

LESSON PRACTICE

List all the possible pairs of factors and tell whether the given number is prime or composite. The first two have been done for you.

1. 16 $1 \times 16, 2 \times 8, 4 \times 4$

   <u>composite</u>

2. 13 1×13

   <u>prime</u>

3. 9   ___ × ___

   ___ × ___

   _____

4. 24   ___ × ___

   ___ × ___

   ___ × ___

   ___ × ___

   _____

5. 7   ___ × ___

   _____

6. 15   ___ × ___

   ___ × ___

   _____

7. 19   ___ × ___

   _____

8. 6   ___ × ___

   ___ × ___

   _____

Practice multiplying by 12 quickly.

9. $5 \times 12 =$ ___

10. $10 \times 12 =$ ___

LESSON PRACTICE 29A

11. $2 \times 12 = \underline{}$ 12. $9 \times 12 = \underline{}$

13. $4 \times 12 = \underline{}$ 14. $7 \times 12 = \underline{}$

15. $12 \times 12 = \underline{}$ 16. $1 \times 12 = \underline{}$

17. $8 \times 12 = \underline{}$

18. Sarah bought three dozen eggs. How many eggs did she buy?

19. Richard is six feet tall. How many inches tall is he?

20. How many months old was Drew when he celebrated his eleventh birthday?

Challenge: Practice skip counting by 12 all the way to 12×12.

$\underline{}$, 24, $\underline{}$, $\underline{}$, 60, $\underline{}$, $\underline{}$, $\underline{}$, $\underline{}$, $\underline{}$, 132, 144

LESSON PRACTICE 29B

List all the possible pairs of factors and tell whether the given number is prime or composite.

1. 2   ___ × ___

   _____

2. 10   ___ × ___

   ___ × ___

   _____

3. 17   ___ × ___

   _____

4. 22   ___ × ___

   ___ × ___

   _____

5. 8   ___ × ___

   ___ × ___

   _____

6. 3   ___ × ___

   _____

7. 12   ___ × ___

   ___ × ___

   ___ × ___

   _____

8. 21   ___ × ___

   ___ × ___

   _____

Practice multiplying by 12 quickly.

9. $12 \times 12 =$ ___

10. $8 \times 12 =$ ___

LESSON PRACTICE 29B

11. 6 × 12 = _____

12. 3 × 12 = _____

13. 11 × 12 = _____

14. 4 × 12 = _____

15. 9 × 12 = _____

16. 2 × 12 = _____

17. 10 × 12 = _____

18. How many inches long is a foot ruler?

19. Riley lived in New Hampshire for five years. For how many months did she live in New Hampshire?

20. Joel bought seven dozen cookies. How many people can he treat if each person gets one cookie?

Challenge: Practice skip counting by 12 all the way to 12 × 12.

_____, _____, 36, _____, _____, _____, _____, _____, _____, 120, _____, 144

LESSON PRACTICE 29C

List all the possible pairs of factors and tell whether the given number is prime or composite.

1. 14   ___ × ___
   ___ × ___
   _____

2. 18   ___ × ___
   ___ × ___
   ___ × ___
   _____

3. 5   ___ × ___
   _____

4. 4   ___ × ___
   ___ × ___
   _____

5. 11   ___ × ___
   _____

6. 20   ___ × ___
   ___ × ___
   ___ × ___
   _____

7. 12   ___ × ___
   ___ × ___
   ___ × ___
   _____

8. 23   ___ × ___
   _____

LESSON PRACTICE 29C

Practice multiplying by 12 quickly.

9. $2 \times 12 = $ _____

10. $7 \times 12 = $ _____

11. $11 \times 12 = $ _____

12. $1 \times 12 = $ _____

13. $6 \times 12 = $ _____

14. $5 \times 12 = $ _____

15. $3 \times 12 = $ _____

16. $9 \times 12 = $ _____

17. $12 \times 12 = $ _____

18. Jeff has to wait four years until he is old enough to join the Boy Scouts. For how many months must Jeff wait?

19. Mrs. Clark ordered ten dozen pencils for the Sunday School classes. How many pencils did she order?

20. Carrie bought eight feet of ribbon for her project. How many inches of ribbon does she have?

Challenge: Practice skip counting by 12 all the way to 12×12.

12, ____, ____, 48, ____, ____, ____, ____, 108, ____, ____, ____

SYSTEMATIC REVIEW 29D

Find all the possible pairs of factors and tell whether the given number is prime or composite.

1. 6   ___ × ___
       ___ × ___
       _____

2. 15 ___ × ___
       ___ × ___
       _____

3. 7   ___ × ___
       _____

Fill in the blanks.

4. 6 ft = ___ in

5. 3 years = ___ months

6. 8 dozen = ___ eggs

Multiply.

7.   4 5
   × 3 3

8. | 4 | 0 | 8 | 2 |
    × | | 2 | 3 |

9. | 1 | 4 | 9 | 9 |
    × | 7 | 7 | 0 |

SYSTEMATIC REVIEW 29D

Add or subtract.

10. 943
    −650

11. 321
    +279

12. 476
    +813

13. Write in words: 76,893,420 _____

    _____

14. Crystal makes greeting cards. She can make four cards in one hour. She worked on cards 21 hours last week. How many cards did she make last week?

15. If Crystal sold all the cards she made last week for $3 each, how much did she earn?

16. Greg has 163 feet of fishing line on a spool. How many inches of line does he have?

17. There are 3,600 seconds in an hour and 168 hours in a week. How many seconds are in a week?

18. Bo Peep found 463 of her sheep at the library and the other 584 of them at the mall. How many sheep did Bo Peep find?

SYSTEMATIC REVIEW

29E

Find all the possible pairs of factors and tell whether the given number is prime or composite.

1. 13   ___ × ___

   _____

2. 16   ___ × ___

   ___ × ___

   ___ × ___

   _____

3. 19   ___ × ___

   _____

Fill in the blanks.

4. 9 ft = ___ in

5. 12 years = ___ months

6. 4 dozen = ___ bagels

Multiply.

7.
```
  1 8 3
× 6 4 4
```

8.
```
  8 7 1 4
×     6 8
```

9.
```
  2 4 0 8
× 7 1 6
```

SYSTEMATIC REVIEW 29E

Add or subtract.

10. 521
    +765

11. 214
    −108

12. 257
    +463

13. Write in place-value notation: 102,500,760 _____

14. Floyd counted the change in his pocket. He has eight nickels and eight dimes. How many cents does he have?

15. Jean flew along in her space ship at 3,500 miles an hour. How much distance did she cover in seven hours?

16. Emily wants to put lace around the edge of a bedspread she made. If the bedspread measures eight feet by six feet, how many inches of lace does she need to buy?

17. Kate was paid $3 an hour for each child she watched. If she watched 2 children for 7 hours, how much was she paid? Multiply two ways to find the answer.

    (__ × __) × __ = _____   __ × (__ × __) = _____

18. Stephanie's keyboard has 26 keys for the alphabet and 76 other keys. Eight of the keys don't work. How many of the keys on Stephanie's keyboard are working?

SYSTEMATIC REVIEW 29F

Find all the possible pairs of factors and tell whether the given number is prime or composite.

1. 10   ___ × ___
         ___ × ___
         _____

2. 17   ___ × ___
         _____

3. 22   ___ × ___
         ___ × ___
         _____

Fill in the blanks.

4. 11 ft = ___ in

5. 10 years = ___ months

6. 5 doz = ___ pencils

Multiply.

7. 964 × 205

8. 3572 × 12

9. 6873 × 343

SYSTEMATIC REVIEW 29F

Add or subtract.

10. 987
    −732

11. 135
    +279

12. 862
    +345

13. Write in standard notation: 200,000,000 + 60,000,000 + 4,000,000 + 500,000 + 10,000 _____

14. Kimberly had 13 jars of grape jelly and 12 jars of raspberry jelly. She used 7 jars of jelly to make sandwiches for her party. How many jars of jelly did she have left over?

15. Alison bought a four-pound bag of apples. How many ounces of apples does she have?

16. Jill has $5 to spend on ice cream cones for her class. Each cone costs 25¢, and there are 19 students in her class. Does she have enough money to get everyone a cone? Will she have enough left to get herself one?

17. Roxanne leads 10 tours a day through the museum. She works 6 days a week. How many tours will she lead in 10 weeks? Multiply two ways to find the answer.

    (__ × __) × __ = _____   __ × (__ × __) = _____

18. One day the museum had 1,547 visitors. If each one paid $5 admission, how much money was made that day?

APPLICATION AND ENRICHMENT

29G

Any shape with four sides is a *quadrilateral*. "Quad" means four, and "lateral" means side.

A *rectangle* is a special quadrilateral with four "square corners." The square corners are called right angles.

A *square* is a special rectangle with all four sides the same length.

Look at the shapes below and follow the directions.

1. Draw a red line around all the quadrilaterals.

2. Color the rectangles blue. There are two of them.

3. One of the rectangles is a square. Draw a green line under it.

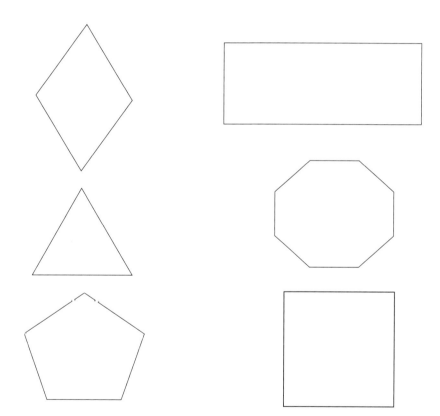

4. Draw two quadrilaterals that are not rectangles or squares.

APPLICATION AND ENRICHMENT 29G

A *rhombus* is a special kind of quadrilateral that has four equal sides. A square is a rhombus, but not all rhombuses are squares because some rhombuses do not have square corners (right angles).

This quadrilateral is a rhombus and a square.
It has four equal sides and four square corners.

This quadrilateral is a rhombus.
It is not a square or a rectangle because it does not have square corners.

This shape is a quadrilateral.
It is not a rhombus because it does not have four equal sides.

This shape is a quadrilateral.
It is a rectangle because it has square corners. It is not a rhombus or a square because the sides are not all the same length.

Circle the names that go with each shape. Most shapes will have more than one name circled.

1. **quadrilateral   rhombus   rectangle   square**

2. **quadrilateral   rhombus   rectangle   square**

3. **quadrilateral   rhombus   rectangle   square**

4. **quadrilateral   rhombus   rectangle   square**

5. Draw a quadrilateral that is neither a rhombus, a rectangle, nor a square.

LESSON PRACTICE 30A

Fill in the blanks.

1. 1 mile = _____ feet

2. 1 ton = _____ pounds

Multiply and fill in the blanks. The first one has been done for you.

3. 3 miles = __15,840__ feet

```
   5,280
   ×   3
   ²
   15640
   15840
```

4. 6 tons = _____ pounds

5. 8 miles = _____ feet

6. 11 tons = _____ pounds

7. 10 miles = _____ feet

8. 4 tons = _____ pounds

LESSON PRACTICE 30A

9. Joanne's new car weighs two tons. How many pounds does it weigh?

10. Walter read the sign next to the bridge: "Weight limit 5,000 pounds." Should Walter cross the bridge if his truck weighs three tons?

11. It is two miles from my house to the grocery store. If I decide to walk, how many feet will I walk?

12. Since there are 12 inches in a foot and 5,280 feet in a mile, how many inches are in a mile?

LESSON PRACTICE 30B

Fill in the blanks.

1. 1 mile = _____ feet
2. 1 ton = _____ pounds

Multiply and fill in the blanks.

3. 9 miles = _____ feet
4. 10 tons = _____ pounds

5. 11 miles = _____ feet
6. 5 tons = _____ pounds

7. 16 miles = _____ feet
8. 35 tons = _____ pounds

LESSON PRACTICE 30B

9. Jane-Alice ran five miles. How many feet did she run?

10. If Chuckie loads his refrigerated trailer truck with 22 tons of ice cream, how many pounds of ice cream will he be hauling in his truck?

11. The mayor wants to have flowers planted along one side of a street in his town. The street is four miles long, and the flowers are to be planted one foot apart. How many plants are needed?

12. A tomato juice factory bought seven tons of tomatoes. How many pounds of tomatoes did the factory receive?

LESSON PRACTICE 30C

Fill in the blanks.

1. 1 mile = _____ feet

2. 1 ton = _____ pounds

Multiply and fill in the blanks.

3. 6 miles = _____ feet

4. 9 tons = _____ pounds

5. 13 miles = _____ feet

6. 12 tons = _____ pounds

7. 30 miles = _____ feet

8. 123 tons = _____ pounds

LESSON PRACTICE 30C

9. A zookeeper ordered 13 tons of hay for his animals. How many pounds should be delivered?

10. Amos had his truck loaded with seven tons of coal. How many pounds will he be hauling?

11. Sarah walked six miles, and Megan hiked 30,000 feet. Who had the longer walk?

12. It is 25 miles from Elm Grove to Pine City. How many feet are between the two cities?

SYSTEMATIC REVIEW

30D

Multiply and fill in the blanks.

1. 8 miles = _____ feet
2. 4 tons = _____ pounds

Find all the possible pairs of factors. Tell whether the given number is prime or composite.

3. 3   ___ × ___
     _____

4. 21   ___ × ___
        ___ × ___
        _____

5. 8   ___ × ___
       ___ × ___
       _____

6. 11   ___ × ___
        _____

Fill in the blanks.

7. 7 quarts = _____ pints
8. 5 nickels = _____ cents
9. 10 dimes = _____ cents
10. 20 yards = _____ feet

Multiply.

11.   5 6 3
    ×   2 4 8

12.   8 6 5 7
    ×   1 5

SYSTEMATIC REVIEW 30D

13.
$$\begin{array}{r} 6\,2\,1\,4 \\ \times\ 5\,7\,2 \\ \hline \end{array}$$

Write <, >, or = in the oval.

14. 11 - 6 ◯ 2 × 4

15. 7 × 8 ◯ 6 × 9

16. 6 × 4 ◯ 3 × 8

17. A railroad car can carry 72 tons. How many pounds can the railroad car carry?

18. Evan filled 47 quart jars with water from the Bay of Fundy and sold them to Darren at a dollar a pint. How much did Darren have to pay Evan?

19. Ooboo the Martian has 11 hands with 11 fingers on each hand. How many fingers does Ooboo the Martian have in all?

20. Roy had a tree that was 14 feet tall. He cut off the top 5 feet to use for a Christmas tree. Since then, his tree has grown another 12 feet. How tall is the tree now?

SYSTEMATIC REVIEW

30E

Multiply and fill in the blanks.

1. 10 miles = _____ feet
2. 8 tons = _____ pounds

Find all the possible pairs of factors. Tell whether the given number is prime or composite.

3. 4   ___ × ___
       ___ × ___
       _____

4. 18 ___ × ___
       ___ × ___
       ___ × ___
       _____

5. 23 ___ × ___
       _____

6. 14 ___ × ___
       ___ × ___
       _____

Fill in the blanks.

7. 9 Tbsp = _____ tsp
8. 7 gallons = _____ quarts
9. $4 = _____ quarters
10. 8 gallons = _____ pints

Multiply.

11. 452 × 71

12. 1372 × 81

SYSTEMATIC REVIEW 30E

13. 4 9 1 2
   × 1 3 1

Write <, >, or = in the oval.

14. 3 × 12 ◯ 4 × 9

15. 50 − 1 ◯ 8 × 6

16. 4 × 18 ◯ 9 × 9

17. A horse galloped for 3 miles before it stopped. If the rancher walked after it for 21,000 feet, did he catch the horse?

18. Jacob's truck weighs 12 tons. It is hauling 13 tons of gravel. What does the truck and its load weigh in pounds?

19. Max bought a rectangular piece of land that measured 613 feet by 466 feet. How many square feet of land did he buy?

20. Craig bought a table that cost $275 and four chairs that cost $68 apiece. How much did he spend in all?

    Craig had $700 to spend. How much money does he have left after buying the table and chairs?

SYSTEMATIC REVIEW 30F

Multiply and fill in the blanks.

1. 12 miles = _____ feet
2. 11 tons = _____ pounds

Find all the possible pairs of factors. Tell whether the given number is prime or composite.

3. 5   ___ × ___
      _____

4. 20 ___ × ___
       ___ × ___
       ___ × ___
       _____

5. 16 ___ × ___
       ___ × ___
       ___ × ___
       _____

6. 19 ___ × ___
       _____

Fill in the blanks.

7. 11 quarters = _____ ¢
8. 10 pounds = ___ ounces
9. 5 feet = _____ inches
10. 12 doz. eggs = ___ eggs

Multiply.

11. 678 × 125

12. 1563 × 64

SYSTEMATIC REVIEW 30F

13.
```
  | 6 | 4 | 7 | 3
  | × | 2 | 1 | 0
```

Write <, >, or = in the oval.

14. 6 × 7 ◯ 4 × 10

15. 9 × 8 ◯ 100 − 30

16. 13 + 14 ◯ 7 × 4

17. Marie drove 33 miles before her car broke down. Round the number of feet in a mile to the nearest thousand. Estimate how many feet she traveled.

18. What is the greatest number of pounds that a vehicle could weigh and still cross a bridge with a five-ton weight limit?

19. Mother made a fish pond for her garden. It is a square that measures three feet on each side. What is the perimeter of the pond?

    What is the area?

20. How many inches long is each side of the pond (#19)?

    Find the area of the pond in square inches.

APPLICATION AND ENRICHMENT

30G

You should already know how to tell time. This activity will help you add and subtract time on a clock. Remember that there are five minutes between each number on the clock. Skip count by five to find the new times.

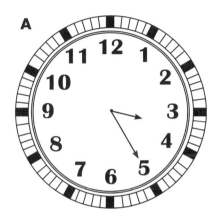

What time is it? _____

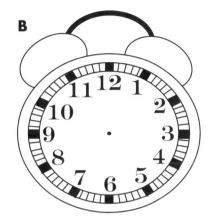

Add 15 minutes to the time on clock A. Draw the new time on clock B.

What time is on clock B? _____

What time is it? _____

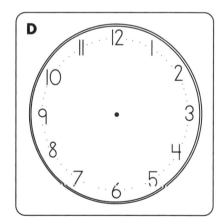

Add 40 minutes to the time on clock C. Draw the new time on clock D.

What time is on clock D? _____

APPLICATION AND ENRICHMENT 30G

Here are more times to add. They are not multiples of five. You will have to add instead of counting by five. For example, the time is 6:14. Bill wants to know what time it will be 42 minutes from now.

```
  6 : 14
+   : 42
-------
  6 : 56
```

When adding times, some problems may result in a number greater than 60 in the minutes place. The student may be able to determine the final answer by moving around a clock face. Regrouping 60 minutes to the hours place will be taught at a later level.

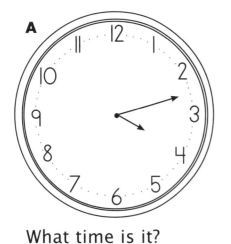

What time is it? _____

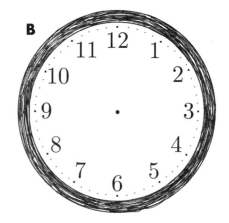

Add 23 minutes to the time on clock A. Draw the new time on clock B.

What time is on clock B? _____

What time is it? _____

Add 16 minutes to the time on clock C. Draw the new time on clock D.

What time is on clock D? _____

APPLICATION AND ENRICHMENT 30G

You can subtract times that are less than 60 minutes in the same way in which you added them. For example, the time is 6:20. What time was it 10 minutes ago?

$$\begin{array}{r} 6:20 \\ -:10 \\ \hline 6:10 \end{array}$$

Add or subtract the times to answer the word problems. You may use a clock drawing or an actual clock to see if your answers make sense.

1. Laurel went to sleep at 8:35, but Emily Jane woke her up 15 minutes later. At what time did Laurel wake up?

2. Stephen arrived home at 5:55. Mom said that she had expected him to get home 45 minutes before that. At what time had Mom expected Stephen to get home?

3. A pie went into the oven at 3:10. It is supposed to bake for 35 minutes. At what time should the pie come out of the oven?

4. Rebekah finished the job at 7:37. That was 17 minutes after she had planned to finish it. At what time had Rebekah planned to finish the job?

Students who have done the application and enrichment pages should be familiar with line and bar graphs, as well as pictographs.

Another kind of graph is called a *line plot*. On a line plot, marks are made above a line to show the number of each item. For example, Bob has five pieces of string. One piece is 5 inches long, three pieces are each 3 inches long, and two pieces are each 2 inches long. The line plot looks like this.

Lengths of String

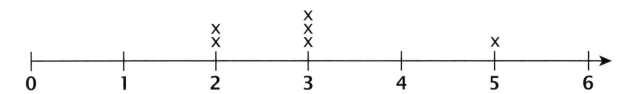

The line plot below is actually a ruler with the fraction lengths marked.

Choose a group of small objects to measure. They should be different lengths. You could measure crayons, pencils, small toys, or even leaves, sticks, or rocks that you have collected outside. Measure each object on the line plot and put a small mark above the correct length for each one. Remember to give the graph a title.

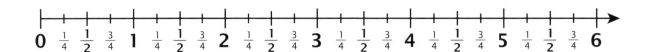

Use your line plot to answer the questions.

1. What measurement did you get most often? _____

2. Do any measurements have just one mark? _____

APPLICATION AND ENRICHMENT

If you wanted to eat one half of a treat, you would cut it into two equal pieces and eat one of the pieces. You can use numbers to write one half. Look at the first square. The 2 under the line tells us the square is divided into 2 equal parts. We call that number the ***denominator***. The 1 above the line shows that you ate 1 part of the treat. We call that number the ***numerator***. The shaded part of the square represents one half of the square. Any fraction with a numerator of one is a ***unit fraction***.

Now look at the second square. The 3 under the line tells us that the square is divided into 3 equal parts. The 2 above the line tells us that we ate 2 parts of the treat. The shaded part of the square represents two thirds of the square.

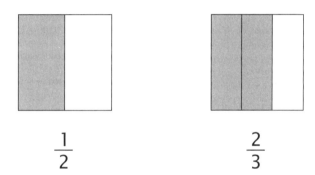

1. Divide each square into equal parts to match the number in the denominator of each fraction. Shade or color the number of parts shown by the numerator. Drawings may have one part or more than one part shaded.

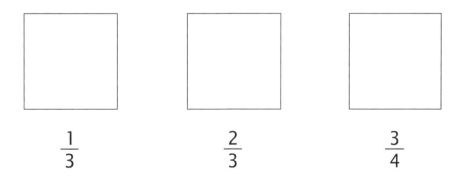

2. If two fractions have the same denominator, you can use a drawing to see which one represents the larger share. Sam ate 1/5 of the cake, and Tom ate 3/5 of the cake. Shade or color the drawings to show what part of the cake each one ate. Put <, >, or = in the middle to show which person ate the larger piece of cake.

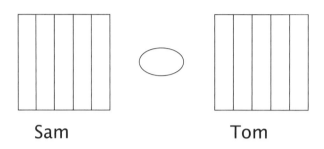

Sam   Tom

You can show a fraction with a number line, which looks like a ruler. Even when something is not shared, you can use a fraction to describe it. Study the drawings and fill in the blanks.

3. The square is divided into _____ equal parts. Two parts are shaded to show the fraction ___ . The line is divided into _____ equal parts. The number line shows us that the fraction is ___ .

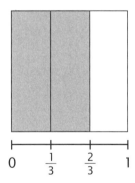

4. This square is one whole. There is only one part, so the denominator is ____. We choose the one part, so the numerator is ____. The fraction 1/1 has the value as 1.

5. The whole amount is ____, and it is not being shared or divided, so there is only ____ part. The fraction 3/1 has the same value as 3.

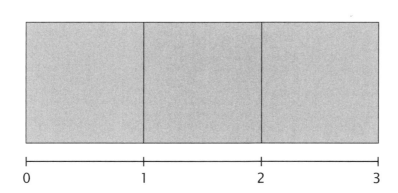

APPLICATION AND ENRICHMENT A1

APPLICATION AND ENRICHMENT

Gamma has focused on teaching students about United States customary measurement. Metric measures are taught in detail in *Zeta*, but students should begin to become familiar with them now.

Centimeters are included on most rulers used in the United States. One hundred centimeters equal one meter. One metric unit used to measure volume is the *liter,* which is slightly more than a quart. A liter is made up of 1,000 *milliliters.* One of the green Math-U-See unit blocks holds approximately one milliliter of water, which has a mass of about one *gram.* A small paper clip also has a mass of about one gram. (Mass is measured in the metric system, instead of weight.) One thousand grams make one *kilogram.* The best way to become familiar with metric measures at this level is to observe labels and real-life applications.

Here are some suggested activities.

A. Collect a number of small objects of different lengths, such as pencils or crayons. Measure and record the length of each one, using centimeters. If you wish, use a line plot to record your answers. As students become more familiar with centimeters, ask them to estimate lengths before measuring the object.

B. A one-liter soda bottle holds one liter of liquid. Use it to compare with other containers to see if they hold more or less than a liter. Have students check their conclusions by filling containers with water and emptying them into an empty soda bottle. (A funnel will make this job easier.)

C. A one-liter bottle of soda has a mass of about one kilogram. Many items in the pantry are labeled with their mass in grams. Have students make a collection of packages that they estimate will have a mass of one kilogram. Add the numbers of grams given on each label to see how close the estimate was. (1,000 g = 1 kg)

APPLICATION AND ENRICHMENT B1

Here are some word problems that use metric terms for measures.

1. Katy bought five bottles of orange juice. Each bottle held two liters of juice. How many liters of juice does she have?

2. Hannah used to have a mass of 27 kilograms. If she increases by five kilograms, what is her new mass?

3. Erin has a collection of 342 paper clips. About how many grams of mass are in the paper clips?

4. Nate is in Canada, where gasoline is sold by the liter. If a liter is about the same as a quart, how many liters should Nate buy to get about a gallon of gasoline?

5. Should Colleen use grams or kilograms to find the mass of a pencil?

   Would grams or kilograms be better to find the mass of a truck?

6. Each side of a square measures 25 centimeters. What is the perimeter of the square?

Symbols and Tables

SYMBOLS

=	equals
+	plus
–	minus
×	times/multiply
·	times
( )( )	times
¢	cents
$	dollars
'	foot
"	inch
<	less than
>	greater than

TIME

60 seconds = 1 minute
60 minutes = 1 hour
1 week = 7 days
1 year = 365 days (366 in a leap year)
1 year = 52 weeks
1 year = 12 months
1 decade = 10 years
1 century = 100 years

MONEY

1 penny = 1 cent (1¢)
1 nickel = 5 cents (5¢)
1 dime = 10 cents (10¢)
1 quarter = 25 cents (25¢)
1 dollar = 100 cents (100¢ or $1.00)
1 dollar = 4 quarters

PLACE-VALUE NOTATION

931,452 = 900,000 + 30,000 + 1,000 + 400 + 50 + 2

LABELS FOR PARTS OF PROBLEMS

Addition

```
  25   addend
 +16   addend
  41   sum
```

Multiplication

```
  3 3   multiplicand (factor)
×   5   multiplier (factor)
1 6 5   product
```

Subtraction

```
  4 5   minuend
– 2 2   subtrahend
  2 3   difference
```

MEASUREMENT

1 quart (qt) = 2 pints (pt)
1 gallon (gal) = 8 pints (pt)
1 gallon (gal) = 4 quarts (qt)
1 tablespoon (Tbsp) = 3 teaspoons (tsp)
1 foot (ft) = 12 inches (in)
1 yard (yd) = 3 feet (ft)
1 mile (mi) = 5,280 feet (ft)
1 pound (lb) = 16 ounces (oz)
1 ton = 2,000 pounds (lb)
1 dozen = 12

Glossary

A-C

area - the measure of the space covered by a plane shape, expressed in square units

Associative Property - a property that states that the way terms are grouped in an addition expression does not affect the result

century - one hundred years

Commutative Property - a property that states that the order in which numbers are added does not affect the result

composite number - a number with more than two factors

D-E

decade - ten years

denominator - the bottom number in a fraction, which shows the number of parts in the whole

dimension - a measurement in a particular direction (length, width, height, depth)

equation - a mathematical statement that uses an equal sign to show that two expressions have the same value

estimate - a close approximation of an actual value

even number - any number that can be evenly divided by two

F-I

factor - (n) a whole number that multiplies with another to form a product; (v) to find the factors of a given product

fraction - a number indicating part of a whole

hexagon - a polygon with six sides

inequality - a mathematical statement showing that two expressions have different values

J-O

multiplicand - in multiplication, the factor that is being repeated

multiplier - in multiplication, the number that indicates how many times the other factor is being repeated

numerator - the top number in a fraction, which shows the number of parts being considered

octagon - a polygon with eight sides

odd number - any number that cannot be evenly divided by two

P-R

pentagon - a polygon with five sides

perimeter - the distance around a polygon

place value - the position of a digit which indicates its assigned value

place-value notation - a way of writing numbers that shows the place value of each digit

prime number - a number that has only two factors: one and itself

product - the result when numbers are multiplied

quadrilateral - a polygon with four sides

rectangle - a quadrilateral with two pairs of opposite parallel sides and four right angles

regrouping - composing or decomposing groups of ten when adding or subtracting

right angle - an angle measuring 90 degrees

rounding - replacing a number with another that has approximately the same value but is easier to use

S-Z

skip counting - counting forward or backward by multiples of a number other than one

square - a quadrilateral in which the four sides are perpendicular and congruent

triangle - a polygon with three straight sides

unit - the place in a place-value system representing numbers less than the base

unknown - a specific quantity that has not yet been determined, usually represented by a letter